The Akashic Experience

Finished NOV'11

The Akashic Experience

Science and the Cosmic Memory Field

Ervin Laszlo

Inner Traditions
Rochester, Vermont

Inner Traditions
One Park Street
Rochester, Vermont 05767
www.InnerTraditions.com

Library of Congress Cataloging-in-Publication Data
The Akashic experience : science and the cosmic memory field / [edited by] Ervin Laszlo.
p. cm.
Includes bibliographical references and index.
Summary: "Firsthand testimonies by 20 leaders in culture and science of their interactions with the Akashic field"—Provided by publisher.
ISBN 978-1-59477-298-6
1. Akashic records. 2. Subconciousness—Miscellanea. 3. Parapsychology and science. I. Laszlo, Ervin, 1932–
BF1045.A44A33 2009
133.8—dc22

2008045917

Printed and bound in the United States at Lake Book Manufacturing

10 9 8 7 6 5 4 3 2

Text design and layout by Priscilla Baker
This book was typeset in Sabon, with Caslon and Futura used as display typefaces

To send correspondence to the author of this book, mail a first-class letter to the author c/o Inner Traditions • Bear & Company, One Park Street, Rochester, VT 05767, and we will forward the communication.

Contents

PART TWO

WORKING WITH THE EXPERIENCE

PART THREE

RESEARCHING THE EXPERIENCE

INTRODUCTION

The Akashic Experience

WHAT IT IS AND WHAT IT MEANS

WHAT IS AN AKASHIC EXPERIENCE?

This volume is dedicated to the exploration of a fundamental yet in the modern world widely neglected aspect of life and consciousness: the "Akashic experience." It contains twenty firsthand reports on this experience by serious, widely known, and highly credible individuals. The accounts make fascinating reading. Before getting down to reading them, however, readers may ask, just what *is* an "Akashic experience"? The question is to the point and merits a prompt answer.

To give a basic definition of the Akashic experience—unlike giving a scientific explanation of it—is not difficult. An Akashic experience is a real, lived experience that conveys a thought, an image, or an intuition that was not, and very likely could not have been, transmitted by our senses either at the time it happened or at anytime beforehand—at least not in our current lifetime. In a popular, though overused and misused formulation, the Akashic experience is a lived experience in the extra- or non-sensory mode.

The Akashic experience comes in many sizes, forms, and flavors, to all kinds of people, and all its varieties convey information on the real world—the world beyond the brain and the body. The experience ranges from artistic visualizations and creative insights to nonlocal healings, near-death experiences, after-death communications, and personal past life recollections. Notwithstanding the great variety in which it occurs, the Akashic experience has strikingly uniform features. Whatever else it may contain, the Akashic experience conveys the sense that the experiencing subject is not separate from the objects of his or her experience—the sense that "I, the experiencing subject, am linked in subtle but real ways to other people and to nature." In deeper experiences of this kind there is a sense that "the cosmos and I are one."

In its many variations, the kind of experience reported in this book suggests that it comes from somewhere beyond our brain and body, and that the information on which it is based is conserved somewhere beyond our brain and body. The Akashic experience gives clear testimony that we are connected to an information and memory field objectively present in nature. In my book *Science and the Akashic Field,* I have given a detailed rationale for naming this astonishing, but in traditional cultures long recognized, reality the *Akashic field.*[1] Here, a preliminary look at cutting-edge science and its rediscovery of this remarkable facet of ancient Indian philosophy will help to explain this name and the nature of the experience to which it refers.

SCIENCE AND THE AKASHIC EXPERIENCE

Science is currently undergoing a fundamental paradigm shift. The currently dominant paradigm of separate material things connected by mechanistic relations of cause and effect is failing; there are ever more things and processes it cannot account for. Classical science's conception of the universe has turned out to be flawed. The primary "stuff" of the universe is energy and not matter, and space is neither empty nor passive—it's filled with virtual energies and information. The universe is an evolving integral system, staggeringly coherent and interconnected.

Leading scientists are discovering a deeper dimension of the universe, a dimension variously called *physical space-time, hyperspace, holofield, implicate order,* or *nuether.* This dimension is associated with the mysterious virtual-energy sea misleadingly called "quantum vacuum" (misleading because this deep dimension is not part of the quantum world but underlies it; and it's not a vacuum but a plenum: it's filled, not empty, space). The unified vacuum—effectively a cosmic plenum—carries the famous zero-point field (ZPF), and grand-unified and super-grand-unified theories ascribe all the fields and forces of nature to it. It is the unified field sought by Einstein near the end of his life.

As scientists now realize, the unified vacuum—now widely known as the unified field—is the originating ground as well as the ultimate destination of all the things that arise and evolve in space and time. In the fiery cosmic birthing of the big bang, pairs of particles and antiparticles sprang from the unified field, and they continue to spring forth in particle accelerators and stellar processes and wherever energies of an extremely high level are produced. In the final collapse of black holes the degenerate remnants of surviving particles die back into this field—perhaps to reemerge again as pairs of particles and antiparticles in the birth of a new universe.

The unified field may have an even more fundamental role. Our universe is not the only universe there is; cosmologists speak of multiple universes arising in a vaster and possibly infinite meta-universe or *metaverse*. The unified field can be assumed to persist even as universes emerge and die back—it's the "stage" on which the repeating drama of cosmic birthing and rebirthing takes place. It's the cradle and the grave of all universes, our universe included.

In regard to its cosmic role and encompassing reality, the unified field is a rediscovery of the ancient concept of Akasha. In ancient India the Sanskrit word *Akasha* meant "cosmic sky," similar to our concept of space. But Akasha referred not only to space in the modern sense but also and above all to the higher spheres of life and existence. The Hindu seers believed that all things arise from, and re-descend into, the cosmic source they called Akasha. Akasha was seen as the first and most fundamental of the five elements—the others were *vata* (air), *agni* (fire), *ap* (water), and *prithivi* (earth). Akasha was said to embrace the properties of all five elements. And it was believed to conserve the traces of everything that ever happened in space and in time. Akasha is the enduring memory of the cosmos: it is the "Akashic Record."

In his classic *Raja Yoga,* Swami Vivekananda described the ancient concept of Akasha like this:

> The whole universe is composed of two materials, one of which they call Akasha [the other is Prana, an energizing force]. Akasha is the omnipresent, all-penetrating existence. Everything that has form, everything that is the result of combination, is evolved out of this Akasha. It is the Akasha that becomes the air, that becomes the liquids, that becomes the solids; it is the Akasha that becomes the sun, the earth, the moon, the stars, the comets; it is the Akasha that becomes the human body, the animal body, the plants, every form that we see, everything that can be sensed, everything that exists. . . .
>
> At the beginning of creation there is only this Akasha. At the end of the cycle the solid, the liquids, and the gases all melt into the Akasha again, and the next creation similarly proceeds out of this Akasha.

Akasha, Vivekananda explained, is so subtle that in itself it's beyond ordinary perception. But when it has taken form, we can perceive it. It's the "real" world that surrounds us.

About a hundred years ago the maverick genius Nikola Tesla revived this idea. He spoke of an "original medium" that fills space and compared it to Akasha, the light-carrying ether. In his unpublished 1907 paper "Man's greatest achievement," Tesla wrote that this original medium, a kind of force field, becomes matter when Prana, or cosmic energy, acts on it, and when the action ceases, matter vanishes and returns to Akasha. Because this medium fills all of space, everything that happens in space can be referred to it.

This insight was not accepted in the science community at the time it was articulated. In the first decade of the twentieth century, physicists embraced Einstein's mathematically elaborated theory of relativity in which a four-dimensional space-time is the ground of reality; they refused to consider anything like an "ether" that would fill space (the search for a unified field that presumably underlies space-time came later). In the absence of matter, space was considered a vacuum. Tesla's insight fell into oblivion. A hundred years later, it is rediscovered.

Today, the concept of an underlying fundamental substrate or dimension in the universe is generally accepted, and the narrow materialism that reigned for more than a century is increasingly abandoned. It has been found that "matter" is a rarity in the cosmos: particles that reflect light and exert gravitation are only 4 percent of the substances that make up this universe. The rest is dark matter and dark energy. Space is a superdense sea of fluctuating energies, and not only of energies but also of information. As physicist John Wheeler remarked, the most fundamental feature of the universe is information—other physical quantities are more like incidentals. Information is present throughout space and time, and it is present at the same time everywhere.

The unified field is a space-filling medium that underlies the manifest things and processes of the universe. It's a complex and fundamental medium. It carries the universal fields: the electromagnetic, the gravitational, and the strong and the weak nuclear fields. It carries the ZPF, the field of zero-point energies. And it's also the element of the cosmos that records, conserves, and conveys *information*. In the latter guise it's the Akashic field, the rediscovered ancient concept of Akasha. A lived connection to this field is the hallmark of the Akashic experience.

THE CURRENT RISE OF THE AKASHIC EXPERIENCE

There is more to human experience than we had thought. Classical empiricism, the long-dominant philosophy of the Anglo-Saxon world, proclaimed that there can be nothing in the mind that was not first in the eye. But classical empiricism was mistaken. Our perception is not limited to wave-propagations in the electromagnetic field and in the air, and to our bodily sensations. We are connected to the world in ways that are deeper and more wide-ranging than our eyes and ears can testify. Recognizing this fact is important—it changes everything we know about the world and about ourselves. It gives us a whole new *Weltanschauung.*

Knowing, or perhaps just feeling and intuiting, that we are connected to each other and to the world in more ways than through our senses is not really new: it's as old as human culture and consciousness. Traditional, so-called primitive people knew that they were connected to each other and to the cosmos; they lived their connections and made active use of them. Shamans and medicine people tuned themselves to a spontaneous apprehension of a deeper reality through rigorous initiation and training; they derived their vision from these nonsensory experiences. The founders of the world's great religions obtained prophetic insights from their connection to a wider reality—even if, in the course of the centuries, their followers embraced the letter of their founders' insight and neglected its substance.

The altered states of consciousness in which our connections become transparent to consciousness have been known and prized by the great majority of the world's cultures. They were known as *samadhi* in the disciplines of yoga, *moksha* in Hinduism, *satori* in Zen, *fana* in Sufism, and *ruach hakodesh* in kabbalah. Their counterpart in Christian mysticism was *unio mystica,* the mystical union of the experiencing subject with the universe.

But today the Western world takes as real only that which is actually to hand—which is "manifest." People dismiss ideas of a wider sphere of reality and consider experiences of it as mere fantasy. Because what modern people see is constrained by what they believe they *can* see, everything that's not conveyed to the mind by the eye and ear is missing from the modern view of the world. Experiences of a deeper or higher sphere of reality are confined to the subconscious regions of the mind and recognized only in esthetic, mystical, or religious exaltation; in love; and in sexual union. In the everyday context the intuitions conveyed by such experiences are ascribed to the unfathomable intuition of artists, poets, prophets, and gurus.

However, modern people's dismissed and repressed experience of a deeper or higher domain of reality is resurfacing in our day. The emerging cultures of peace, solidarity, and respect for nature no longer believe that all we can experience of each other and the cosmos must come to us through "five slits in the tower"—they know that we can open the roof to the sky. They realize that human existence is not bound by our ego and our skin, that we can have contact and communication beyond the reach of our eye and ear. More and more people are having Akashic experiences.

The Akashic field—the information and memory components of the unified field—is not mere theory: it's a part of the real world. And, as the reports in this volume testify, it's an *experienceable* part of the real world. Access to the Akashic field—the Akashic experience—is a genuine and indeed fundamental element of human experience: as Edgar Mitchell suggests in this book, we should regard it not as our *sixth* sense, but as our *first*—it is actually our most basic sense.

THE MEANING OF THE AKASHIC EXPERIENCE FOR OUR LIFE

The recognition that the Akashic experience is a real and fundamental part of human experience has unparalleled importance for our time. When more people grasp the fact that they can have, and are perhaps already having, Akashic experiences, they will open their mind to them, and the experiences will occur more and more frequently, and to more and more people. A more evolved consciousness will spread in the world. People will shift from the ego-centered skin-enclosed consciousness of the modern age to the interconnected transpersonal consciousness foretold by thinkers such as Sri Aurobindo, Jean Gebser, Richard Bucke, Rudolf Steiner, Stanislav Grof, Don Beck, Ken Wilber, and Eckhart Tolle, among a growing number of others.

The Akashic experience testifies that we are subtly yet effectively linked with each other, with nature, and with the cosmos. It inspires solidarity, love, empathy, and a sense of responsibility for each other and the environment. These are ineluctable elements of the mind-set we need to pull out of the global crisis that threatens our world and to create peace and sustainability on this perilously ravaged planet.

A BRIEF GUIDE TO THE CONTENTS OF THIS BOOK

Part 1 of this volume contains firsthand reports that recount the lived Akashic experience of the writers. The reports testify that, regardless of in what form it occurs, an Akashic experience proves life- and mind-transforming. Sometimes it begins at an early age, triggered by a trauma or serious health problem. In other cases it comes about later in life, marking a new stage in the development of the individual. In this part the accent is not on a theoretical explanation of the experience; the reports serve first of all to document that the experience occurs and that it transforms the thinking and even the life of those to whom it occurs.

Part 2 includes a more practical dimension: it outlines how an Akashic experience can find productive use in everyday life, as well as in education, business, healing, and art. The experience, the reports tell us, can open fresh vistas in education and can provide valuable guidance in business. It suggests effective approaches to health and healing and is a source of inspiration for painters, composers, and all creative people.

In part 3, while the reports continue to describe the experiences lived or witnessed by the writers, the experiences are subjected to systematic scrutiny: they become objects of scientific research. The varieties of Akashic experience studied by the writers embrace the many forms of "psi" (parapsychological phenomena) as well as near-death, out-of-body, and after-death communication and past life experiences. The tools used for researching these experiences include innovative methods in psychotherapy and experimental parapsychology as well as clinical observation and statistical analysis. These endeavors break fresh ground, because for the most part an Akashic experience cannot be explained by the concepts and methods of conventional science—it doesn't fit its materialist and reductionist paradigm.

Part 4 undertakes an overall review and assessment of the Akashic phenomenon. The writers, who have had an Akashic experience themselves, know that it's a real experience with notable healing capabilities and life-transforming effects. They have devoted their lives to asking how we can best make use of this experience for our own well-being and the well-being of all people in the human community.

The path traced by the four parts takes the reader from *firsthand accounts* of life- and mind-transforming experiences to the *practical application* of the experiences, then to the *analysis of the conditions and effects* of the experiences, and finally to the *assessment of the meaning and life significance* of the experiences. This trajectory leads logically to Summing Up. Here this writer endeavors to give a bona fide scientific explanation of the Akashic experience and thus breach the long-standing divide between this deep domain of human experience and the recognized fields of scientific research.

Part One

LIVING THE EXPERIENCE

ONE

Journey Home

MY LIFE-TRANSFORMING AKASHIC EXPERIENCE

C. J. Martes

C. J. Martes, healer and author, has helped clients in more than forty countries for nearly twenty years. In 2004 she developed Akashic Field Therapy (AFT), an integral method of quantum healing that helps individuals identify and then remove subconscious negative patterns and beliefs at the mental, physical, and spiritual level. Her work blends A-field (Akashic Field) Theory, Behavioral and Integral Psychology, Vibrational Medicine, and Western science.

During the course of my life I have been privileged to experience some quite amazing things. These experiences have shaped my life. They made me a better person. They stretched me to the limits of my consciousness. They made me aware that there is a whole world "out there," in which there is much more than meets the eye.

I came into this world wide awake it would seem. My mother told me that her fondest memory was of holding me the day I was born. My head was covered with dark brown hair. I was staring at her with a pair of the largest, bluest eyes she had ever seen. She said they were as big as saucers.

That childhood image seems very indicative of how I would experience the world. I was a curious child, who had to know everything. I would sit in my mother's lap for hours, asking her about this or that. She would patiently explain things of the world to me.

My earliest Akashic experience happened when I was about seven years old. I seemed to have an extra awareness of the world around me. I remember watching people and immediately knowing a lot of things about them. I would see things in my mind before they happened. At first I thought that all people experienced life that way. I later found out they didn't.

I grew up in a small town in Kansas and attended a Catholic school. There was a small cathedral where I went to church three days a week. This cathedral holds many fond memories of a spiritual nature. I recall sneaking away from class or the playground to go in the church by myself. I was harshly scolded a time or two by the nuns for my disappearances, but that never deterred me from going.

I remember standing there, dwarfed by the towering stained-glass window of the cathedral, with the sun casting rainbow colors all along the wall and across the floor. My blue eyes were transfixed, and I tuned out all the other sounds around me. Beautiful. Breathtaking. I was in awe.

The church was a special place, quiet and expansive. I loved everything I felt when I was there all by myself. I would walk in quietly and sit in the very first pew. I thought intently about God and the angels. I used to question God about the meaning of life. I talked at length with God and heard a powerful voice respond with love and understanding. Through these dialogues, I had a sense about the essence of life, that nothing in it was really material. I grew up knowing that life was only the way it seems to be because we perceive it to be that way.

In that church I felt connected to the divine. For a time, I even wanted to enter the sisterhood because the calling I felt was such a strong force in my life. I believed there was a reason God whispered messages directly to me that no one else could hear. It felt like I was being called home. I talked about it often to the adults that would listen. I'm not sure they knew what to say to me, so most of them would say very little. They would simply smile and nod or change the subject.

I remember so much from those early days. Life was a challenging mystery and I tried earnestly to figure it out. In school I attended the mandatory catechism classes. I remember asking the priest all kinds of awkward questions, like: Where is God? Why are we here? Am I like God? Most of my early questions about such things were dismissed readily by whomever I asked, but I nevertheless kept asking them. I have quite persistently questioned the reality of everything my whole life.

A MESSAGE FROM BEYOND

When I was twelve, I had an otherworldly experience. My family had just made an abrupt move from the small town where I grew up. My life as I knew it was in upheaval. I missed the town and my home very much. I had to adjust to new friends and a new school. I missed the cathedral most of all. I remember having a difficult time transitioning. Everything felt awkward and uncertain.

One evening I went to bed very tired. I remember thinking about my hometown and the old church while I was trying to fall asleep. Around one in the morning I awoke abruptly. I sat up in bed somewhat groggy. When my eyes focused I saw the translucent figure of a woman standing at the foot of my bed. I recognized her as my great-grandmother. She had been dead at that time for more than five years. Time seemed to stand still while I looked at her. Her presence didn't frighten me. In fact, a feeling of great peace and love spread through me. It reminded me of the feelings I had when I sat in the cathedral talking with God.

She began to speak to me without moving her lips. She told me that she had a message to give me about my grandfather. She said he was sick but no one knew it yet. She said that she wanted me to know what was happening since he and I were so close.

My grandfather lived in the small town I had just left. He and I had a special bond, and I cherished our time together. We would take long walks in the afternoons. He and I would play card games all the time. He beat me at Gin Rummy more times that I can count.

The news great-grandmother was giving me was startling. A deep sadness filled me. As if trying to soften the blow, she assured me that he would not suffer but that he would eventually pass away from the ailment he had. After lingering for a few more moments, she disappeared.

I sat there for a few moments, trying to take in what had happened. Then I lay awake for quite a while thinking about what she had said. I wondered what was making him sick. The next day, I told my mother what had happened. She said it was a strange dream for a young girl to have. As was usual whenever I brought up this kind of subject, we did not have a dialogue about it.

I knew in my heart that my grandfather was sick and the experience was not just a dream. I asked to call him later that evening. We spoke for a while and he seemed just fine to me. I tried to put the experience in the back of my mind. This worked for some time.

Several months later I found out what was wrong with my grandfather. I came home from school one day in the afternoon. When I first saw my mother she had a serious look on her face. She told me to sit down and then said, "Your grandfather has Alzheimer's disease."

I questioned her since I didn't know what that was. My mother, who is a nurse, explained it to me. She didn't remember what had happened several months earlier, but I did. The experience in my bedroom came flooding back to me. I knew what my great-grandmother had been trying to tell me. My grandfather had shown early signs of Alzheimer's disease within that several-month period, but it hadn't yet been recognized.

My grandfather died several years later. Because we lived so far away, I didn't get to see him as much as I would have liked. Still, I knew he passed away long before I received the phone call from my mother.

I was in bed preparing to sleep one evening. The house was completely silent. I was under the covers and turned facing the wall. Suddenly, I felt a presence enter my room. This presence hung in the air. It was tangible and vibratory. I looked over my shoulder and turned around in bed. There stood my grandfather. His peaceful figure stood there smiling at me. I looked at him for a long time. He looked so happy and well, as if he'd never been sick.

Without saying anything to me, he walked over to my bed and sat down next to me. He took my hand in his. I could feel some pressure and a deep tingling sensation in my hand. At that moment I could see and feel every moment we had spent together. The memories flipped through my mind like photographs, one by one.

In just a few minutes, I vividly experienced every beautiful moment I had ever spent with my grandfather. Tears filled my eyes. I remember his beautiful smile and the love in his eyes while this was happening.

When our time of remembering was over, he stood up. He said, "Please tell Mom I love her and I'm all right." I nodded and said I would. There was a bright light all around him. The light intensified for a moment and then he disappeared. I was so grateful that I had been given the chance to say good-bye to him that evening.

It was nearly a year later that I was able to deliver the message from my grandfather to my grandmother. It was Thanksgiving. The entire family was busy getting things ready. My grandmother was in town, staying two blocks away at my aunt's house. I decided to walk over to see her before dinner.

When I got to my aunt's house it was very loud inside. My three cousins were running in and out of the house. My aunt was busy finishing up in the kitchen before heading to my house with all the goodies. My grandmother was seated at the kitchen table. I sat down next to her and said hello. After a few minutes, everyone seemed to disappear and the house became entirely silent. I had not planned that day to talk to my grandmother about the experience with my grandfather. But a knowing feeling came over me. I took the sudden silence as the opportunity to speak to her about it.

I remember being nervous. I wasn't sure how she'd respond to what I had to tell her. I overcame my anxiety by reminding myself that I had made a promise to my grandfather. I carefully explained to her what had happened after he died. Then I told her the message he gave me.

Tears welled up in her eyes. She grabbed my hand and gave it a sincere squeeze. She looked me deep in the eyes and asked me if I had the gift. I nodded. She later told me that I'd given her the best Christmas present that anyone ever could have. I was elated with her response and very much relieved. She believed me.

Soon after that experience, my childhood experience of God and the cathedral gave way to teenage experiences. As I grew up, my childhood experiences began to seem like distant echoes that were far, far away. I didn't question life so much. I kept many opinions to myself. I never seemed to see things before they happened anymore. The world of Akashic experiences was gone from my life. Or so it seemed.

HEALING, NOT HURTING

The many experiences I had in childhood didn't prepare me for what happened in my midtwenties. To recount this experience I need to go back several months.

I met my husband by chance one day while visiting an old friend. My husband would say it was love at first sight. Our new relationship came on the heels of several failed relationships that had been emotionally abusive. I had been through a lot of personal turmoil prior to meeting him and had many unhealed wounds. Even so, there was an immediate connection and I trusted him.

The thing I remember so well after getting to know him was that our hearts were the same. We looked at life through the same set of eyes. We spent all our time together. We were happy, so lucky to find each other. In the early days of our relationship we talked for hours about the important things and found we had similar views. There was a powerful connection. We were both coming out of failed marriages, but we believed love would carry us out of the past and into a shared future.

Eventually we decided to move in together. My personal decline began soon after that. I barely noticed it at first. Melancholy thoughts seemed to surface even when good things were happening. At first, this was quite subtle; I could barely notice it.

We had a good relationship. Anyone looking at our situation from the outside would have said that things were finally looking up for both of us. Yet I was increasingly saddened by the smallest things. Slowly the color drained from my life. I felt my heart slipping away bit by bit. Within months the sadness that had been merely a blip on the radar screen overcame me entirely.

I began experiencing with vivid detail all the previous negative moments in my life. It was like reliving my traumatic experiences one by one. Each memory was endowed with an intensity that I cannot fully describe. I became despondent and literally stuck in the past. There was so much good happening in my life, yet I couldn't be consoled by it. Nothing gave me happiness or hope.

Soon the depression was far reaching and consumed every waking moment. No matter what I did, I felt worse instead of better. My days were filled with nothing but the pain of the past. The pain I felt inside was intense. It was like each event was a still frame that my mind reviewed one by one. I tried my best to sort it out. It was exhausting.

Eventually I had very little energy for the basic things in life. I spent days in bed with no desire or motivation to rise even to take care of my basic needs. There was little anyone in my family could do to help me.

Looking back, I feel sorry for my husband, who was helpless in the face of all I was going through. He tried desperately to cheer me up. He brought me flowers, thoughtful cards, and talked to me for hours on end. I'm sure in his own way he felt as bad as I did.

I became not only depressed but also very angry. I couldn't understand why I was literally falling apart when things were actually going in a positive direction. It seemed like the ultimate irony. Here I had someone that loved me for myself, yet I was miserable and couldn't enjoy it. I even thought once or twice that God must be punishing me for some unknown sin.

I felt sorry for myself between my bouts of anger. Several months passed as I fought myself endlessly. It was like being haunted by a world of desperate shadows. My shadow side was in full evidence every day. It was thick and ominous. It was the only thing I could see. I was sick on the inside. Sick and tired of feeling the way I did every day.

Then one day I was going through my usual routine of lying in bed and spending short periods sitting in one of the easy chairs in our bedroom. I remember sitting there, mentally sorting through the previous months of life in a dark oblivion. I felt bruised from head to toe. I lacked the will to continue with the present situation. I was weary. I thought that my situation would never end. I was alone in the room, but I started to speak out loud to no one in particular.

"I'm so tired of hurting. I just can't do this anymore. Why is this happening to me?"

The room was silent. No answer came.

"I feel so alone. I feel like nothing matters. Why do I feel this way?"

Still no answer.

"Someone take this pain from me. I'm tired of hurting."

Then I heard a soft clear voice. A curious thought came to me from the air.

"You're not hurting."

"What do you mean, not hurting!" I said angrily to the voice I heard.

"You're not hurting."

I sat in silence for several moments. Then I heard the voice again, quite clearly. It was as if another person was sitting next to me. Although the voice came from nowhere, it didn't alarm me.

"You're not hurting. You are healing."

I considered what I had heard very carefully. Healing had never entered my mind before. I tried it on for size. I thought about what it meant to be healing instead of hurting. What if the voice were right?

I spoke again: "If I'm healing then I can't do it alone." At that very moment, I raised my hands toward the ceiling inviting anyone who might help me. I opened myself up for the first time in a long time. I had no idea that one single gesture would produce the incredible events that followed.

What happened next was the most tangible Akashic experience I've ever had; all that had happened in my childhood paled by comparison. It felt like a discernable electric charge was shooting out of the air around me. What was conveyed to me in milliseconds was unlike anything I had ever felt before—I was tingling from head to toe. It was the consummate spiritual experience in one enormous flash of light.

I felt the room fill up all around me. I felt an intense love, pure and unconditional. It was an experience also of nonduality, an experience of perfect oneness. The hair on my arms stood up. I was rapidly surrounded by a bright light. The tiny room became crowded with angelic beings. There were twelve of them, each comprised of different-colored light. They formed a circle around me. I could hardly believe what I was seeing.

As I watched the image unfolding in my room, I realized that I had never been alone in everything I had been experiencing. These beautiful loving beings had always been there with me. They were trying to help me. My emotions had prevented me from feeling their presence.

Everything became instantly clear. I could see the difference between healing and hurting. I knew what it feels like to be hurting, so I tried to imagine what healing would feel like. While looking at everything as an experience of healing, my perception shifted effortlessly from a perception of utter pain to one of hopeful reflection. A calm sensation filled my body. I relaxed for the first time in months. Healing was feasible. I could see healing as a process in which I was participating. Instead of endless suffering I could see that there would be an end to what I was experiencing emotionally. At that moment all the fighting I had been engaged in against myself left me. There wasn't anything to fight anymore. My inner conflict subsided.

I call that day my awakening. Things that had been dormant seemed to come alive inside me. All the gifts I had as a child came rushing back. I could see so many things. All my senses were heightened and alive. I could perceive energy all around me. No matter what I was doing, the angelic beings were always with me. I no longer felt alone. For a long time, I could communicate with them. It was like the veil between worlds opened up and I could see directly through it.

I learned more about life in that moment than I had in all the previous twenty-five years. I had been liberated from my self-created prison by a single word. I learned how to heal myself and to let go of the past. A different idea entered my consciousness that day and set me free. I realized that my perception of the situation had kept me hopelessly stuck—it had been the key to everything. My months' long ordeal had been due to how I had looked at what I was going through.

AT HOME AGAIN

After that day my whole world shifted dramatically. I could enjoy life again. I resumed my daily activities almost immediately. I was given a great gift that day, and I wanted to share it with others. I told everyone I could what had happened. I suppose some believed me and others didn't. But that no longer mattered to me. The experience recalled my early days in the cathedral. I had finally come back to the comfort of my childhood. I had finally come home.

The experience changed my life in countless ways. I can see the rich tapestry of human existence for the beautiful thing it is. I can believe in the power of the divine again. I can challenge myself to become more than I was before. I decided shortly after that experience to use my gifts to help others just as I had been helped that day. I want others to see what I've seen and help them to overcome their own challenges.

Life can often be a journey full of the unexpected. It can sometimes be difficult to face when there's uncertainty in the heart. Now the uncertainty I used to feel has left me. Each day I can meet life's challenges with more confidence because I know that I am never alone.

These Akashic experiences have affected me forever. I have touched the divine and the divine has touched me. There's always more to life than meets the eye; there's always more to see in any situation. Most of all, I possess the comforting knowledge that no matter where I may find myself, I can always get back home.

TWO

Experiences of Infinite Consciousness

Swami Kriyananda (J. Donald Walters)

Swami Kriyananda (J. Donald Walters) is founder of Ananda Sangha, a worldwide spiritual cooperative community organization dedicated to the teachings of Paramhansa Yogananda with branches in Assisi, Italy, and California. He is a member of the Club of Budapest and has lectured widely in the USA, Europe, India, and other countries. Kriyananda has written eighty-five books and composed more than 400 pieces of music. He lives in Gurgaon, India.

The Akashic Records might seem to be some heavenly office room filled with file cabinets. *Akasha*, however, means "space." The "records" implied are intrinsic to the oneness of infinite consciousness. They are called so because it's possible to access any specific part of omniscience you wish. Omniscience is not a "ball of wax," available only if you will accept the whole thing.

Let me give an example. I was with my guru, Paramahansa Yogananda, when he finished writing his commentaries on the Bhagavad Gita. He commented to me, "Now I understand why my guru [Swami Sri Yukteswar] never wanted me to read other commentaries on the Gita.

He didn't want me influenced by other people's opinions. Before writing these commentaries, I tuned in to Beda Byasa [the author of the Gita, Veda Vyasa; my Guru used the Bengali pronunciation of the name], and asked him to write it through me. These commentaries express Byasa's consciousness, channeled through me."

Possible? Even conceivable? Let me tell another story. One time, Yogananda wanted to relate an episode from the life of an Indian Bengali saint named Chaitanya. He said, "I will let him tell this story through me." The story that emerged had a beautiful immediacy.

Many people would say that in both cases he was speaking only poetically. I, who lived with him, believe, quite to the contrary, that he spoke the literal truth.

His poem, "Samadhi," in *Autobiography of a Yogi,* a description of his experience of cosmic consciousness, contains a line that says, "thoughts of all men, past, present, to come . . ." All these exist in those "Akashic Records"—not "filed away," but accessible by simply tuning in to the appropriate "ray" of Infinite Consciousness.

I don't compare myself in spiritual development with my guru, but I can offer one or two of my own much more ordinary experiences in accessing those "records." In college I passed a Greek exam for which I had done almost no studying (the subject really *was* "Greek" to me!) by telling myself firmly, "I'm a Greek." Suddenly I found myself attuned to that ray of consciousness that results in the Greek language. Only two people in the class passed that exam, which turned out to be an unusually difficult one. I was one of them.

I speak several languages and have always found it easiest to learn a language by first attuning myself to the vibrations of consciousness that produced it. I try from the first to *think* in that language. I rehearse myself in the accent and try to get it as right as possible. I tell myself, "This, now, is how I *want* to think: in *this* language." Thus, I eliminate from my mind all sense of its foreignness and never translate the phrases mentally from the English, with the right English verbal sequences, but accept those new sequences as right and natural.

MUSIC FROM THE AKASHIC FIELD

I have found the same to be true in other fields: when writing music, for example. I knew almost nothing of this subject when the inspiration first came to me to try my hand at composition. As far as I knew, there were only two rules: that parallel fifths (whatever that meant) are supposed to be a bad thing; and that, generally speaking, the base line should move in an opposite direction to the melody.

The inspiration to write music came because it seemed to me a good way to reach those people who might find philosophical truths less palatable when stated in a book or lecture. In other words, I wasn't thinking of music itself as having any particular capacity to convey states of consciousness. However, as I offered a philosophical thought up to Infinite Consciousness, a melody suitable to that thought would come to me almost effortlessly, in an instant. Gradually I realized that music itself actually expresses states of consciousness: it's not only a pleasant (or unpleasant, as the case may be!) kind of self-expression. I remembered, for example, one popular song from the 1920s that had been banned from the airwaves because too many people, after hearing it, had committed suicide. The song's title? *Gloomy Sunday.*

As a practitioner of meditation it came naturally to me to be calm. While composing I found that when I didn't identify personally with the emotion I wanted to express, the underlying *spiritual* vibration of that feeling came to me without effort, for the mere asking. What came was a sort of "soul's-eye view" of a human emotion, which provided a much clearer understanding without involving me in any mood. Thus, what was perhaps the saddest of my songs came to me one day when I was serene and happy. It delighted me, actually, to the point of making me laugh as the words and music appeared!

Many books teach us to affirm our own personal ability to excel in whatever we want to do: as a salesman, for example, or in public speaking, or as a rising executive in a large business corporation. I learned the opposite from my guru, however: that pride is "the death of wisdom." The secret is to get the ego out of the way. Difficult? In one way, yes,

but in another way, actually simple. What problem can there be in surrendering one's identity to Infinity? I learned to affirm not "I can do it!" but, "It's there to be done; let me access that particular 'ray' of knowledge which shines from Infinity within whatever requires to be done."

In music, consequently, when I asked in meditation to be inspired with the right words and melody, the inspiration came almost effortlessly: the melody instantly, and the words with a fluency that amazed me. Whole songs, both words and melody, would come while I was walking in San Francisco's Golden Gate Park or driving across the Oakland Bay Bridge.

One day I was driving to an appointment, hampered by heavy rain, and for a few minutes was mentally grousing at this obstacle. Suddenly the thought came to me, "I can do nothing about this rain, but why not at least have 'good weather' inside?" Instantly, the melody and the first words of a song came to me: "There's joy in the heavens, / a smile on the mountains, / and melody sings everywhere! / The flowers are all laughing / to welcome the morning, / Your soul is as free as the air!" The song went on to emphasize the importance of rising above all self-shackling desires. Even as I drove, I wrote what I could on a little paper beside me, trying still to drive with relative safety. At the earliest convenience, after reaching my destination, I wrote down the whole song. It has been received everywhere with smiles of keen appreciation. "Painless philosophy," I call it!

More and more, as I pursued this avocation, I found that melodies themselves, presented with the right chords and rhythms, reached people instantly with truths that didn't even need words to convey their clear meaning.

Later, I found that I could also tune in to a country and instantly receive a melody that was suited to the essential consciousness of that country.

One day I remarked on this strange fact to two friends. I said, "Let me just go to the piano and see if I can't play something Japanese." I knew next to nothing about Japanese music, but as I placed my fingers on the keys I asked for a suitable melody. Instantly the notes came, in as little time as it took me to play them. Another friend of mine who had lived for many years in Japan exclaimed, "Why, that's perfect for what you've called it!" Its name? "Cherry Blossoms in Kyoto."

Again, I knew nothing of Persian music, but I wanted a melody to go with the *Rubaiyat of Omar Khayyam,* for which my guru had written a deeply spiritual commentary. One morning as I awoke from deep sleep, the melody was already in my mind. I had dreamed it. Years later I played it to someone from Iran, and he exclaimed, "Why, that's Persian!" Many melodies have come to me, indeed, in sleep.

The most sophisticated form of composition is, I've been told, the string quartet. I know very little of this kind of music except that I enjoy listening to it. One day, however, I decided to try my hand at writing such a quartet. Banishing from my mind any thought that this was even a challenge, I simply asked for guidance. My formula always is simply to say, "I can't do it, but You, God, can do anything!" I also deeply believe that everything we do should express our own philosophy of life. In writing this quartet, I decided that I, as the founder of several cooperative communities, should give all four instruments a chance to express themselves, instead of having three of them do little but accompany the melody, played by the first violin. (So you see, I do have to know clearly what it is I am asking for.) I can't possibly explain how it happened, but suddenly I heard those instruments playing in gay counterpoint to one another, soaring and dipping as if dancing together.

A professional classical violinist, who not long ago assisted in recording this quartet, told me, "The second movement is as good as anything by Beethoven!" I don't know what to think. I'm not the composer and take no credit for the music. I do like that quartet—in fact, I love it—but I simply wrote out what I received.

ATTUNING TO INFINITE CONSCIOUSNESS

I have learned that all knowledge is available to us. We don't have to create it; we have only to access it. Simply ask in the right way—not with pride in your accomplishment, but with an open heart. I don't even mean to ask humbly, in the sense of being self-deprecating. Don't think about yourself at all, nor about your ability or lack of it. Concentrate, rather, on attuning yourself to Infinite Consciousness and ask for guidance in what you want to do. It's delightful, fun, and deeply inspiring to work and let yourself be used in this way.

I worked with my guru on editing his commentaries on the Bhagavad Gita. He told me that part of my work would be editing. As things worked out, however, it became necessary for me to plow in different fields from my fellow disciples. I therefore had no access to his manuscripts. I wasn't satisfied, however, with what had been published of that work. It didn't reflect what I felt was its true spirit. Nor did it fully express the content.

One day I thought, "I'm growing old. Who knows if I'll ever be able to do this work, but it absolutely needs to be done." So I asked Yogananda inwardly to guide my efforts and was amazed, after more than fifty years, to find that I could remember, idea for idea, everything he'd said for every stanza. I began the work on October 7, and ended it on December 5: less than two months for a work of nearly 600 pages. Published since then in book form, it has received glowing reviews. Most important, however, where I am concerned, is that I feel it is faithful to what he finished in 1950.

I have found that Infinite Consciousness can also supply material needs and not only the necessary information or inspiration. One day in our Ananda Village near Nevada City, California, we had a fund-raising event for the beautification of what we still whimsically call "downtown Ananda." Requests were made for flowers ($10), bushes ($15), and other small items. There was one request, however, for $3,000 (I believe it was) to cover the cost of an improved driveway. The hope, evidently, was that one of the Ananda businesses would come up with a pledge for this amount. When I saw that request, I thought, "We certainly need an improved road, but who will pledge so much?!" I had no money. Still I thought, "God can supply it." With that thought in mind, I wrote down my name pledging for the entire amount.

Impossible? Of course (as far as I myself was concerned)! But I must have asked in the right way, for one morning a week or so later (the money had been requested within two weeks), I found an envelope thrust under my front door. It was from a former resident of the community who had just come on a visit. The letter said, "My mother passed away several months ago and left me some money. I've long wanted to express my gratitude to you for all you've given me. This is my little way of doing so." Enclosed was a check for $3,000.

This sort of thing has happened many times in my life. I hope that whoever reads these pages will not think I am claiming some special gift for myself. What I want to say is simply that we are *all* surrounded by an ocean of abundance: knowledge, wisdom, ability, opportunities, material plenty. What a pity it is that people close themselves off from that spiritual environment. Keeping their gaze fixed on the ground, they trudge through a life burdened with worries, fears, and self-doubts.

I could continue with many stories of a similar nature. These few, however, should suffice to make my point. There is, in omniscience, all the knowledge and inspiration you need, and, in omnipotence, all the power.

There exists, however, another phenomenon of such an extraordinary nature that I thought to finish this contribution by recounting it.

ASTOUNDING PREDICTIONS

Many years ago (1959) in Patiala (Punjab), India, a son of the Maharaja of Patiala, a student who was taking a course I was giving in Raja Yoga came to me one day at the home of Balkishen Khosla, where I was staying, and asked, "Swamiji, have you ever heard of Bhrigu?"

When I couldn't place the name, he helped me by adding, "Bhrigu is mentioned in the Bhagavad Gita, where Krishna [speaking in the voice of God] says, 'Among saints, I am Bhrigu.'" Of course then I recognized the name. Bhrigu lived in India in very ancient times.

Raja Mrigendra Singh, my visitor, went on to say, "Bhrigu wrote a *sanhita* [scriptural document] predicting the lives of innumerable individuals yet to be born, some of whom are actually living today."

This seemed to me, of course, almost too fantastic. Yet I had already encountered examples of the bizarre and the unusual in that mystical land. To preface what came next, let me recount an ancient Indian tradition, which my "spiritual grandfather" (my guru's guru), Swami Sri

Yukteswar, clarified and, so to speak, "pruned" of inaccuracies that had crept in under the disintegrating influences of time. The tradition concerns four ages, explained by Swami Sri Yukteswar as being brought about by sidereal movements within the galaxy. That system is too complex for explanation here, but it is also related to the Akashic influences.

Sri Yukteswar stated that the earth recently entered *Dwapara Yuga,* the second of those ages, in which human beings will come increasingly to understand that energy is the basic reality of matter. In this Dwapara age also, humans will gain insights into the essentially illusory nature of space. Thus, in the centuries to come we will learn how to travel to other planets and to demolish the sense of spatial distance. This we have accomplished already to some extent, with the invention of the telephone, radio, television, Internet, and air travel.

It is said that in the third of the ascending ages, *Treta Yuga,* humans will develop insight into the essentially illusory nature of time. We will understand that time and space are much more elastic than they have seemed; time itself will be increasingly perceived as a continuum, comparable to a river that, when observed from a bridge, is seen to consist not only of what flows directly under the bridge but also of the water flowing down to the bridge from upstream and away from it downstream. In other words, the future already exists, being the result of flowing influences from the past, and will not change significantly with anything added to the water—perhaps cast into it from the bridge.

Hints of this reality are suggested already even today. They will become so obvious in the third yuga as to be universally accepted. Particularly gifted individuals will be able, beyond the denials of any cynic, to predict specific events far into the future.

Even today, predictions have been made, mostly regarding the lives of individuals but also regarding world events, that have turned out to be startlingly accurate. The knowledge of enlightened sages, moreover, has always shown itself in this respect to be quite extraordinary.

I was told a story, based on the personal experience of someone I knew who had visited a saint in Howrah, West Bengal. He had asked the saint how accurate and how specific a prediction could be. The saint responded by foretelling several completely unexpected events that would occur to him that very afternoon. What he said (and here, I am able only to paraphrase) went something like this: "When you leave here, you will be obliged to take a detour because a crowd will have gathered in the street in front of a burning building. On that detour, you will see an accident on the right side of the street, but it will not impede you, and you will have a safe journey home." The details were not exactly as I've related them here, but what actually occurred was comparable. I was assured that the prediction had come to be fulfilled in every respect.

Yogananda's *Autobiography of a Yogi* contains many predictions of a similar nature. I'd like to emphasize that I myself lived with the author of that book as his close disciple and am fully convinced of its, and of his, veracity.

Back, then, to my own experience with Raja Mrigendra and Bhrigu's text, which was extraordinary. Raja Mrigendra told me that not many miles from where we were, "in the town of Barnala, there is a partial copy of that ancient document, in manuscript form. It contains predictions of the lives of individuals, including many who are living today. I found there a reading for myself. Would you," he continued, "be interested in going there to see whether the *sanhita* contains something about you?"

"Are its predictions only general?" I asked. "Might it say about me, for instance, that I've come from a distance and appear to be interested in spiritual matters?"

"Nothing like that!" he replied confidently. "If it says anything at all, it will be much more specific."

Well, naturally I was intrigued! We went by car the next day to Barnala, a town that in no way suggested mystical wonders, being an assemblage of completely ordinary, somewhat dirty streets and buildings, none of them even interesting. The structure that housed the miraculous document was quite as nondescript as anything in its surroundings. We were the first to arrive, and I was introduced to the custodian, a *brahmin* whose name (if memory serves) was Pundit Bhagat Ram. He welcomed us, showing my friend the deference due to his social position.

Passing lightly over the formalities, a horoscope was cast for the moment when I asked for a reading. The pundit went into an inner room where the stored document was piled on many shelves in bundles. He retrieved a small bundle numbered (I vaguely recall) 54. Opening the bundle, and dividing it into three piles, he kept one himself, gave one to Raja Mrigendra, and handed the third stack of pages to me, telling the two of us to look for a page showing a horoscope similar to the one he had drawn up. We each went carefully through them. It was I who came upon a page that seemed to me similar to his design. It was the right one.

"The readings," Raja Mrigendra had told me, "usually tell a person his last life, his present one, and his next one." My reading began, as he'd predicted, with my previous life. It told me that in that life I had been born in India. My name was Pujar Das. I lived in Karachi (identified by the first letter in the name of that city, and also by its geographical location), was married, and was financially well off. We had no children. There followed a brief description of my life up to the time where my wife and I went on a pilgrimage and came to a desert (probably in Rajasthan), there reaching the place where the ancient sage Kapila (founder of the Sankhya system of philosophy) had once had his ashram. There I met my guru. I resolved to stay there and seek God, sending my wife home. A fair amount of information followed, all of it both interesting and instructive, but too personal for inclusion here. None of it was verifiable, of course, though it's true that in my present life I have felt strangely attracted to living in the desert.

"In the present life," it continued, "he was born in a *mlecha* ['unclean,' an ancient word for Western] country, is well known as a seer of *Ashtanga Yoga* [the teaching of Patanjali], and is traveling and teaching in this country. His name is Kriyananda."

This piece of information brought me up sharply. I was astounded. Kriyananda is a most unusual name, though two or three monks (*sannyasis*) have taken it since I did. Several more people had entered the room by now, and I passed the page around to them to see if they could verify whether this name was indeed written there. They all concurred that it was. The "reading" omitted mention of my next life but made a few predictions for this one that were interesting and hope inspiring, if a little vague.

The fact that it mentioned me by name, however, was itself simply amazing. What it said about this life, also, was more or less accurate, though general. Would I have liked more specificity? I'm not so sure. Sometimes it's more helpful to have a general sense of one's direction than to be burdened with too many details, whether alarming or giving comfort.

What was I to think? The reading closed by saying, "There will be no more readings today." Everyone in the room, accordingly, left with us.

I was fascinated enough by all this to speak about it to friends. It seemed to offer evidence, above all, that there was much more to India's ancient civilization—as I of course already knew to be the case—than cowherds, farmers, and primitive villages. Surely what it suggests, rather, is a legacy of extraordinary wisdom. This was ammunition that would help to substantiate any book or lecture on those ancient teachings.

A few weeks later I was giving lectures and classes in New Delhi, where this new interest led to another segment of the Bhrigu sanhita. Here I received another reading. It stated, "I have already given him a reading in my *Yoga Valli*. That one was according to astrology. This one will be according to the power of yoga." Instead of once again telling me my last life, it went back to an earlier life.

"In the time of Kurukshetra [the historic war described in the ancient epic the *Mahabharata*], he was the ruler of a small state in Bharatavarsha [India]. Fearful of having to support the wrong side in that conflict, he handed over his kingdom to his son and went into the forest for a life of seclusion and meditation. There he took initiation from a guru." The reading went on to describe that man's life, saying that after it, owing to his good deeds, he spent some 700 years in the astral world.

Fascinating! In many ways that subtle region has always seemed more real to me than this physical world, though what remain are strong impressions rather than clear and specific memories. Again, I purposely omit here details of that past life that are personal and not germane to these pages.

What ensued then was even more astounding than the reading in Barnala. "This life," it continued, "is the eighth since that one during the time of Kurukshetra. In the present life he was born in Romania, lived in America [both statements were correct], and his father named him James. [James is in fact my first name, though I was always known by my second name, Donald.] He has two brothers, but no living sister is possible, though one will die in his mother's womb. [My mother admitted to me, after my return to America, that she had had one miscarriage.] After meeting his guru, Yogananda, his name will become Kriyananda. Within two months from the time he receives this reading he will return to his own country, where he will be lovingly received by his (spiritual) brothers and sisters, and will be given [appointed to] a high position."

Interestingly, I was in fact summoned back to America within two months. On my return voyage, while visiting Japan, I received word that Dr. M. W. Lewis, the elderly vice president of my guru's organization, had just left his body. Shortly after my arrival in California I was appointed to replace him as a board member and as the vice president. So far, the accuracy of the reading was uncanny. Until the death of Dr. Lewis, I had already achieved as high a position in the organization as was available to me. The reading went on to make several pleasing predictions, all of which came true.

It then went on to say, "What I have given him so far are the fruits of his good karma. Now I shall relate the fruits of his bad karma. There is a danger to him of sudden, unexpected death. There is also the possibility of obstacles arising in his mission, and in his *sannyas* [his spiritual dedication]."

Not long afterward, on my return voyage to India from America, I bought a Lambretta motor scooter in Italy and had it shipped to India. In the large, high-walled courtyard of the home I used to visit in Old Delhi, I unpacked the Lambretta, sat on it, and turned the key to the "on" position, assuming the machine would then wait for "further instructions." Horrifyingly, the motor was already in gear, and suddenly, without the slightest warning, the scooter took off at full speed across the courtyard toward that high wall. I had no idea what to do. In slightly more than a second I would have to find the solution. Somehow I managed to get the motor out of gear, hastily found and applied the brake, and stopped barely six inches short of what would have been almost certain death.

Not very long afterward, the "obstacles to his mission" that Bhrigu had predicted burst upon me: totally unexpected misunderstandings on the part of my superiors, and summary dismissal from the organization to which I'd completely dedicated my life, with no hope of appeal or reprieve. Finding myself suddenly thrust back into the "world," with neither support nor preparation, I was for several years plunged into a maelstrom of inner turmoil. Fortunately for me, my heart's dedication to seeking God through service to my guru never wavered, and I came safely through this period of very serious testing.

The reading went on to describe a future of fame, success, and good fortune. This one, again, was silent about my next life. Most of the page was devoted to this life, giving details that, over time, proved true. The reading did say one thing about my future, which could be of interest to many people. What it stated was, "In the future in his country, when there will be weeping in every home . . ." and spoke of my role in those trying times. I can easily accept that prediction of "weeping in every home" as a possibility, for I have long believed, and indeed my guru predicted, that a serious depression will come, which will be, he said, "much worse than the depression of the 1930s." To another disciple he said also, "The dollar won't be worth the paper it's printed on." If you take to heart only this much of what I have written, you may be helped by receiving a salutary warning. The solution my guru suggested was that people band together, buy land out in the country, and form small, self-sustaining cooperative communities.

An Indian acquaintance of mine from Los Angeles told me that an Indian friend of his had gone to the Bhrigu sanhita and had been told, "As this is being read out, there will come a thunderclap." Presumably this was to be a sign of the reading's veracity. The sky that day was completely clear of any cloud. Just at that moment, however, there came a loud thunderclap. (The philosophical and scientific implications of this event are simply staggering.)

I suspect science will not adjust easily to the possible existence of Infinite Consciousness and of the awareness, within that consciousness, of every stray thought in every scientist's mind. Krishna in the Bhagavad Gita states, "In the atheist, I am his atheism." Yet modern research is tending more and more to the strong suspicion, if not the actual conclusion, that there is more in "them thar hills" than anyone nowadays has yet guessed, or can even imagine.

WHISPERS FROM ETERNITY

Let me conclude with another true story. Someone once asked Paramhansa Yogananda, "Is it possible for inspiration to be under the control of one's will?"

"Certainly!" replied the master. He was busy at the moment preparing to go out, but he stopped, sat in a chair, and said, "Take down these words." He then dictated:

"O Father, when I was blind I found not a door which led to Thee, but now that Thou hast opened my eyes, I find doors everywhere: through the hearts of flowers, through the voice of friendship, through sweet memories of all lovely experiences. Every gust of my prayer opens an unentered door in the vast temple of Thy presence."

This prayer-poem was included later in Yogananda's published book, *Whispers from Eternity*. A reviewer, in praising the book, wrote, "It contains one poem that I can't resist quoting." He then went on to quote this particular poem.

Everything you have ever wanted to know; every talent you've ever wanted to possess; every satisfaction you've ever wanted to achieve—all these await you already in the Akashic field (Infinite Consciousness, subtler than space itself), surrounding your every thought, hope, ambition, and desire.

psychometry

THREE

Return to Amalfi and the Akashic Home

David Loye

David Loye, psychologist, evolutionary systems scientist, and author, is the founder of Benjamin Franklin Press. The full account of his investigation of past lives, six books of the Darwin Anniversary Cycle, six books of the Moral Transformation Cycle, and an Entertainment and Humor Cycle are among twenty new books published at the Benjamin Franklin Press by Loye.

Something those who have personally experienced the so-called paranormal know is this. Scientists in the future will marvel over how, in its dominant attitude toward this subject, the science of the twentieth century so blindly mirrored the arrogance and hostility of the religion of the Middle Ages toward science itself.

In both cases we see the attempt to call a halt to evolution by shutting the windows into the wider realm of knowing, which Laszlo's bold venture to connect new wisdom with old wisdom now ever more surely reveals as an Akashic field.[1]

In my case, the first window on the Akashic field was my discovery of psychometry. This is where you can hold someone's watch or ring in your hand, put yourself into a quick light trance, and receive images

out of their past, present, or even sometimes, I suspect, their future. Doing this, with generally startling results, I've entertained hundreds of people over the years, thinking as I did so, "How typical: here is this miraculous power, but—at this still imprisoned stage of evolution—we have only learned to use it for trivial entertainment." This first window had opened in the late 1970s while I was moving from Princeton on the staid and skeptical East Coast to the more open-minded wonderland of California and the UCLA School of Medicine. Soon, as I write of in *3,000 Years of Love,* the window of telepathy opened up.[2]

At UCLA I was disgusted to find that psychologist Thelma Moss was being shoved off the faculty. The university was afraid the notoriety of her pioneering work to get at the roots of psychic healing would degrade its image and endanger fund raising. I sat in on a forlorn exploration of telepathy conducted by a little research group she'd left behind, soon became their sponsor, and within a few weeks became one of the inner circle of supersensitives. After a year of what had by then become routine success with telepathy, I plunged on to precognition with the group. Our startling success sparked my writing of *The Sphinx and the Rainbow* (published in 1984 and updated in 2000, in terms of chaos theory, with the new title *An Arrow Through Chaos*).[3]

In the first half of *The Sphinx and the Rainbow* I played it safe, by writing about how we use the brain and mind to read the future in terms of the acceptable canon for science. In the last half, however, I boldly ventured into precognition, which resulted in two great boosts of encouragement. Pioneering brain scientist Karl Pribram praised my theory of how precognition worked in terms of a "hololeap," based on Karl's pivotal but then still controversial holographic brain and mind theory. After reading *Sphinx,* Ervin Laszlo invited me to join the exploration that became the General Evolution Research Group, and which has also taken him deeper and deeper into the Akashic field in book after book.

Back in those years of bread and butter necessity, I was cautiously working to build the safe reputation that might gain me foundation funding and a tenured professorship. But when I read of the brilliant work of Russell Targ and Harold Puthoff in remote viewing, I found that this fascinating new window into the field revealed the possibility of practical uses beyond entertainment.[4] I was able to use remote viewing several times when my children or my wife's children, hundreds or even thousands of miles away, were in possible danger and couldn't be reached. In order to check on them, I simply put myself into a light trance and saw they were safe. At the same time, I was able to confirm the power of this new tool of mind, for I saw what I could not possibly have experienced in any other way. Details of what I had observed were later confirmed by the children, which proved I'd actually "gone there in mind," such as a strange sound I once heard, which turned out to be the exact thump, thump, thump sound of the snow chains on the bus carrying a targeted stepdaughter as it climbed into a blizzard on a mountain.

All this, however, turned out to be only the upstairs windows for the great wide door that opened before me with the venture into past lives I write of in my book *Return to Amalfi*.[5]

THROUGH THE DOOR INTO THE PAST

As it is the direct experience that convinces all but the most hopeless of skeptics, this I will now try to provide by relating some events, just as they unfolded.

I am in the pleasant guesthouse in which Nadya Giusi, a fellow psychologist, works with clients in my hometown, Carmel-by-the-Sea in California's Monterey Peninsula. Nadya is sitting nearby. I am sprawled on the stereotypical couch. I have put myself into a light trance with three deep breaths. I nod to Nadya.

During an earlier session she suggested that I focus on some area of persisting pain in my body in order to seek the past. Another earlier session began with the question, "What do you see when you look down at your feet?" This time, however, as I have demonstrated a facility for reentering other "lives" with minimal inducement, Nadya skips the preliminaries.

"Where are you? What do you see?"

Almost immediately I find myself on the shore of what appears to be a seaport town. I am down close to the water.

"Nearby I can see a dock or wharf, to which a ship of an ancient type is tied. It seems to be a three-master with the high prow and rear of the Middle Ages. I can see the sailors unloading its cargo. Nearby to the left I can see a table at which several men are seated. One has an open book, a ledger, into which he is either jotting things down or checking things off."

"Who are you?"

"I am . . . huh!" I chuckle with surprise at the certainty with which I seem to know that "I am a spice merchant here to receive the consignment of spices the ship has brought in for me from the Orient."

I can't get over it. It's like I'm in a theater eating popcorn and watching a movie at the same time, only this is not like a movie in the sense of something projected on a screen. I am simply there—*in* the movie while eating popcorn.

Most weird of all is that in comparison to what shows on a movie screen this ostensible past life is rounded, full dimensional. It's something of which I am an intimate part, while the "popcorn present life" seems like what physics tells us it is, a jumble of atoms racing through time.

It's as though the dream has become real and the real has become the dream.

"What year is it?" Nadya asks.

"1611," I say without hesitation.

Rather than a vague "somewhere in the Middle Ages," here out of nowhere emerges this precise year!

1611?

This places my experience back in the time of the late Renaissance—which raises the question of why I have picked *this* life. Why this grubby nobody?

If I am just inventing this, why—in such a rich period for creative genius and colorful intrigue—am I supposedly just another one of those billions of us who century after century pass in and out of life, leaving nothing lasting upon this planet or the flow of time?

With all the great figures of the late Renaissance, as well as all the colorful and meaningful occupations of that time to choose from, why did I purportedly find myself to have been a lowly spice merchant in what appeared to be Italy? Why had I not discovered I was Michelangelo lovingly polishing the stone that brought to life the great Pietà of Mary holding the body of the dead Jesus in her arms? Or Leonardo da Vinci putting the twist of abiding ambiguity to the face of the *Mona Lisa*? Or at least a famous poet or explorer in that age of ages for magnificent venturing?

Why on earth would the mind of someone in our time go so far out of the way to select the seemingly meaningless life of an obscure spice merchant for a past life?

"And now what is happening?" Nadya's focusing question comes.

"I have concluded my business and am now going up into the city. And going *up* is very much the experience. There is a steep grade from the waterfront upward into the city. The climb is also made somewhat difficult by the uneven roughness of the stones underfoot."

I find myself wondering why there would be such a steep grade in a seaport town, which would make it very difficult to haul goods up from the ships. It also didn't make sense that there would be so little level land between the waterline and the grade upward. Logically, a seaport should have ample level land for warehousing. Yet there appears to be none here.

"Where are you going?" Nadya has but to ask, and it is as though I am both the actor with the leading role in a movie and the cinematographer tracking myself with a slow zoom as the experience continues to unfold.

I am going home, I realize. I continue to ascend up, up, up the steep grade from the waterfront, the rough stone paving underfoot, into a patchwork of streets that wind off in various directions.

Pursuing one direction off to the right I come upon an area of houses with a specific look. They seem to be two stories or more high, of massive wood and stucco or stone construction.

The wood is dark, forming a framework for the stucco or stone. The wood uprights and crosspieces form the lower story, while the wood beams are crisscrossed for the second and other upper stories. The second story also slightly overlaps the first level.

Now I am approaching what I know to be my home. It has a door partitioned in the middle, so that the top half can be opened while the bottom half remains closed. It also has a latch of a specific type still found in use in old farms into recent years. It has a bar that falls into a slot, which is raised by placing pressure on a thumb piece. It is locked in place by a peg inserted into a hole on the inside, which then prevents anyone from opening the latch from the outside.

I open the latch and go inside. The ceiling is quite low. I have a vague sense of having a wife here, but she doesn't register clearly enough for a report.

"Is there someone of meaning to you in this life you can describe?"

I am aware I have a daughter who means the world to me. She is our only child. She is sixteen. She becomes ill. It's terrible. It has something to do with her throat. She has difficulty breathing. A woman healer whom I like and have confidence in seems to be able to relieve and soothe her to some degree, but my wife insists on bringing in a male doctor whom I rather quickly dislike and have no confidence in. My daughter gets steadily worse. In terrible agony, she dies and I am left with a deep, abiding bitterness and hatred.

Here in Carmel—both inside and outside the "movie"—my emotional response is so strong that I feel like pounding on something.

But I am lying down.

I can do nothing but *feel* the suddenly overwhelming need to pound on something—and be frustrated.

I hate doctors with a passion for their futile quackery. I hate God with a passion for taking her away. Above all, I hate the Church, which is involved in some way. In an age and place when faith in God and the Church is a basic social, political, and economic requirement, my reactions are curious. But the spice merchant who ostensibly is myself is a fervent atheist. Still, this apparently does not hurt my business as I am able to hide these feelings behind a mask of affability and the ways of a small merchant's charm.

Nadya inquires: "Can you find a meaningful event of some kind in this life?"

The setting changes, and I am on a high plateau above the city. It is a picnic area for the people of my class, small merchants, tradesmen, and craftspeople, the bourgeoisie of the time. It is a spot of green above the red tile roofs of the city. We are having a picnic when there is a sudden outcry. A young girl has fallen over the edge of the cliff toward the roofs below.

I am getting agitated. I wet my lips and look around.

"Do you want a drink of water?"

"No." I don't want to risk losing my connection to the scene. "Let's go on."

I rush to the edge of the cliff and look down. The girl has caught hold of a root or stone and is clinging there. If she falls, she will not likely be killed. It is only twelve, or at most twenty feet down. But still she could be grievously hurt, with broken bones.

Quickly I arrange a rescue. Asking those nearby to give me their jackets, shirts, or blouses, I twist each garment and tie them together to form a long thick rope of clothing. This I loop at the end and lower over the side of the cliff to the girl. She clasps it, hangs on as we drag her upward, and so is rescued.

"What next?"

I find myself in a vast lighted chamber. It is a place of authority, the palace or main receiving chamber for the ruler or chief authority for the seaport at that time. I am to be honored for the girl's rescue.

The chamber is of a wondrous size, with the glitter of gold and many candles. The ruler sits on a raised dais, his deputy close at hand. I and my friends are in a little knot of people off to the left about twenty to thirty feet or so.

I am aware that my honoring is to be only a very minor matter for the day, but still it is a great thing for me and my friends. We wait while the deputy reads this and that and the ruler rises and declaims and sits again, and at last it is my turn. The deputy reads a brief account of the rescue from a scroll. The ruler rises and compliments me, hands me the scroll, and the tableau is over.

"What does he look like, this ruler?"

"He is a man in red with heavy eyebrows, debonair, with a wax-pointed mustache and fierce Van Dyke pointed beard. He looks much like the pictures of the great bass Feodor Chaliapin in his role as Mephistopheles, the devil, in Gounod's opera *Faust*."

Further focusing on him, I have the impression he is an arbitrary figure to be feared, not liked, with a certain playful charm dancing over a cold, cold heart. He has a resemblance to a Spanish conquistador—the same faintly contemptuous look and imperious bearing shown in the pictures in history books.

"What else?"

Somewhere, too, there emerges a fleeting memory of my childhood. I am on a ship. It is my father's ship. He is the owner of this and other boats. I love being at sea.

"Now let us advance to the day of your death. What do you see?"

I am surprised to find myself again on the very same green plateau above the city where the rescue of the young girl had taken place earlier. Looking down from the plateau, I am again struck by the sight of the mass of red roof tiles below with the sea in the distance beyond.

I am now about fifty years old, a respected elder among the happy gathering of the small merchant and craftspeople at a picnic, families, children, all having a good time. Everything is going along well until two young men become involved in a loud dispute. They pull out pistols, and everyone is instantly alarmed. Feeling both a sense of my responsibility as an elder and also a sense of confidence in myself as one who can restore peace ("Ah ha," I think as this is happening, "I am again the rescuer as in other past lives and so often in my fantasies"), I walk toward them remonstrating.

Kindly, gently, I urge them to put away their pistols and calm down. But to my astonishment, the young man to my right, about twenty feet away, turns toward me angrily and fires his pistol. To my further astonishment, I realize that I have been shot through the heart. But I can feel nothing—only surprise this could be happening to me. I feel no pain, but—knowing what a shot through the heart means—I realize I am dead. And all I can feel is an overwhelming sense of astonishment; I even feel the irony and the weird humor of the situation. I could easily laugh.

MURDER? WHY, WHERE, OR WHEN?

Did this really happen?

This is the first question that hit me after this experience. Was this really me in a past life? If so, there had to be other lives out there. Out of so many possible lives, why this particular one? And why should it respond to an invitation to "come through" in this unlikely small town at this particular time?

Here, at least, was a murder, raising the old question of who had done it? And why? Or what indeed was it all about?

The damper of practicality comes quickly with the sense that this can be nothing more than fantasy. But the final twist defied the door slam of complete skepticism. I could write off the rest as only invention; it could all be explained as merely another example of the facile invention we display in nighttime dreaming, daydreams, or as writers generating fiction. But to be shot through the heart, feel no pain, know you are dead, and feel like laughing?!

My early years of studying successful and writing unsuccessful short stories, plays, and even novels, made it clear to me that the oddness of my reactions had the ring of something not easily dismissed. A writer—even one of genius—would either go into the agony, show the proverbial all-of-life flashing by in an instant, or tough guy it, clip it short, like Hemingway. But to be shot, feel no pain, know you are dead, and feel like laughing—this had the feel of the weird touch of reality beyond invention.

If there was anything to my experience, where had it all supposedly taken place? I had a date and a rather detailed description of the place but no sense of where this might have been other than the strong feeling it had been in Italy. But it was not at all the Italy with which I had any familiarity.

It was about as far removed in looks and feel as possible from the Italy that my wife and I knew from our visit to the turreted hill towns and the rolling hills of Tuscany, where we'd stayed in Ervin's villa near Pisa. Nor was it anything like inland Rome, or Florence, where we'd gone for the first major conference for our General Evolution Research Group. Puzzled by this odd configuration of a green plateau above red tile roofs with the sea below, I described the place to a fellow venturer into the mysteries of the then new field of chaos theory and evolutionary systems science, Monty Montuori.

"Amalfi," Monty said immediately. "There are seaport towns that look like that along the Amalfi coast. It may even have been Amalfi itself."

RETURN TO AMALFI

Four years passed and gradually the experience faded. Then my wife and partner, Riane Eisler, simultaneously sold the Italian publishing rights to both her increasingly popular earlier book, *The Chalice and the Blade*,[6] and her new one, *Sacred Pleasure*.[7] Nuova Pratiche Editrice in Milan was going to publish *Il Calice e la Spada* and Frassinelli, also in Milan, was going to publish *Il Piecere e Sacro*.

Once again the door into the Akashic field beckoned. And for the first—or second—time I was in Amalfi. I was overcome at first sight, for—as I stood beside the highway high up on the slope in here-and-now reality—I saw precisely what I had seen in that very real earlier reality.

Night was coming on, the breeze now cool, coming down from the mountains, the shadows etching slope and city in brilliant chiaroscuro. There, down the great downslide of rock from the Latari mountains, was the same stair-stepping sequence of green plateaus above red rooftops. There, touched by the after light, was the same warm dotting of ivory buildings. There was the same sea stretching out blue and shadowed but still aflame, endless, into the horizon.

I felt an inner welling up of gratitude, of the sense of sought-after bonding, of fulfillment, with such intensity that it was inescapable, beyond denial. It was like coming home from the war, or from college when young, or reconnecting with one's children after a long absence.

But would everything else I had seen and experienced back in Carmel also be confirmed? I will juxtapose the questions with what I found.

AKASHIC Q & A

1. In terms of the relation of the seafront and possible wharf to the town, was the waterfront area now the same as I had seen it?

I immediately felt this could definitely have been the place. The sweep of the narrow strip of beach was the same, as was the sense of the ascending presence of the town behind and, above all, the breakwater. For there, straight out before me, was the breakwater, or narrow rocky wharf, precisely where the dock or wharf had been in my experience of the 1600s!

But wouldn't whatever had been here 400 years ago have been destroyed by now?

Yes. And of course the favored location for handling ships could have been shifted elsewhere. But then I was struck by the fact that later structures, whatever they are, generally follow the same line in relation to the land or water as earlier structures.

Part of the reason can be a matter of favorable geography, such as the best placement for a wharf in relation to the prevailing tides, winds, and closest proximity to where the goods are to be unloaded from the ships to where they are to go.

Part of the reason can also be a matter of tradition—the conservative feeling that because the old structure was here, so should the new one be. Otherwise, it just won't look right. This could perhaps be the reason above all in Italy, where—in sharp contrast to the raw, new, slam bang tradition for the U.S.—for many centuries the look of things was more important than whether it made money or was maximally convenient. There it was likely that later structures would be located in the same place as earlier structures due to the Italian feeling for aesthetics.

In any case, there the breakwater was, jutting out about two or three hundred feet into the bay. I could almost place the ship I saw so long ago beside it, unloading.

2. Would I find the steep streets I saw? And those hard, rounded, cobblestones I had felt underfoot?

In the piazza at the Amalfi waterfront I found the same stones. Rough and black and of both odd and even shapes set in mortar underfoot. But was this just a patch that would soon give way to asphalt or concrete as I ascended into the town?

As I climbed up street after street in search of what might have been my house or any other memorable detail, the stones were still everywhere underfoot. Indeed, the higher I went, the more marked the presence of the stones became. As John Steinbeck had written of the Amalfi coast's even better-known comparable attraction, Positano: "Its houses climb up an acclivity so steep that it could more appropriately be called a cliff, were it not for the flights of steps that have been built into it."

3. Back then had Amalfi been a notable site for the spice trade?

I had thought of my spice merchant's trade as having been a rather grubby and drab calling. Nowadays the spice rack in the grocery store is small in comparison to other departments. We usually use only two or three ourselves, maybe cinnamon, thyme, dill, plus some mixture, such as Italian spices for pizza. But when I started exploring the history of Amalfi, I found that in the centuries before modern refrigeration, spices were the world's great substitute—particularly in the hot countries around the Mediterranean where food went bad in a hurry.

With the right spices, you could make a delicious meal of vegetables on the verge of becoming a smelly mush or meats just this side of turning green. They also helped preserve cooked foods longer.

Historically this basic functional advantage led to pleasant discoveries as more and more cooks experimented with spices, particularly in Italy and France. For it was found that with them cookery could be turned into an art form with the same assertions and nuances as painting or music.

The more I looked into it, the more I realized that the spice merchant of that time must have had the kind of commercial charisma and social status as a dealer in computers or software enjoys today. Back then the well-stocked and knowledgeable spice merchant would have offered the lure of all the wonders in eating experiences, similar to the current allure of the electronic wonderland of software.

"Its ships sailed all the seas, trading with the countries of the Near and Middle East, transporting spices and silks and the precious woods sought after in the West," Steinbeck had written of Positano. If Positano—with less of a harbor than Amalfi—had been very active in the spice trade, this most certainly would have been a major industry for my old "hometown." Steinbeck specifically identified the sixteenth and seventeenth centuries, in which I was interested, as a time in which spice trading was still of major importance in the area. *A Short History of Amalfi,* which I found in a tourist shop, gave further confirmation, stating that the products exported by Amalfi to the East "were mainly rosewater, iron, wood, conserves, and hand-made products, and the products imported to Italy were spices, perfumes, seasonings, silks, carpets, and gems."

4. Had Amalfi still been functioning as a seaport when supposedly I lived there, not as anyone whom history might have in any way noted or revolved on, but simply and obscurely as an ordinary spice merchant?

This question was underlined by my discovery that in 1073 CE the city was struck by a tidal wave of such immense power that it sank every ship in the harbor and wiped out what was then half the town along the waterfront clear back to the Duomo, forcing the survivors to rebuild higher up. This explained the strange configuration I had noted during my experience when I was in Carmel, which was reaffirmed on site in Amalfi: the steep rise and tough haul up from the edge of the water into the town and the lack of customary land for unloading and warehousing.

My study of the history of the area showed that this devastation had definitely led to Amalfi's decline as a seaport. Could it still have carried on the active trading I had supposedly witnessed in the year 1611?

The prints of local paintings I found in the Amalfi shops made it evident that the answer here was also an unequivocal yes. I discovered at least a dozen different views of Amalfi painted over the centuries. Curiously, all were radically different from one another. The overall look was the same, but the details of what was there and what wasn't there differed. Another difficulty was that none of the old prints were dated, so I had to guess at time periods according to the style of ship and kind of clothes people seemed to be wearing. Most, however, showed ships in the harbor.

Some ships were at sea. Some were anchored nearby or tied to wharfs. Some had been hauled up on the beach, presumably at high tide. In any case, it was clear that Amalfi had continued to serve as a seaport, however minor, well through the sixteenth century into our own time.

5. Might I even find—as impossible as it seemed after such a long time—something still standing that looked like what was then my own house, which I had seen, and had even opened the door and walked into?

Definitely not. The current buildings were of sheer stone flat surfaces.

But somewhere, at an earlier time, I *had* seen the same kind of houses of stone or stucco with wooden beams and crosspieces that I had seen during my Carmel "visit to Amalfi." Searching my memory, I realized that I had seen them years before in Bologna, in northern Italy. Though they had been built in the Middle Ages, many were still standing.

Perhaps the same type of house had been present, if only in patches, in the Amalfi of the early 1600s but had since been destroyed. Or perhaps my conscious mind, seeking a definite image for what might have been an indefinite mental picture of a house, had simply thrown in the stock image it already had for "house in the Middle Ages in Italy." Much research shows this is often how our minds work in reconstructing memories of past events in our present lives.

6. Would I find a door latch of that particular type? The top half of the door that swung open? The peculiarly low ceiling?

Again, no. Now the prevailing style is vertically split doors—which have the look of having been around for a long time. Again, either the horizontal splits were there earlier, or my conscious mind had tossed in a stock image from the memory bank.

7. And now the death of my beloved daughter. What about the situation of her getting better under the care of the woman healer I believed in but dying after my wife called in a male surgeon, and my ferocious hatred for the Church thereafter?

Here the book that had helped to pay our way to Amalfi, my wife's *The Chalice and the Blade,* provided part of the answer. Careful research has shown, as she reports, that the brutal torturing, burning, and hanging of anywhere from 100,000 to nearly a million women who were called witches was the outgrowth of a much more complex process of downgrading women by the Catholic Church of the Middle Ages. Rather than being mad or odd women who supposedly cast spells on people and sexually consorted with the devil, most of these women had been the traditional healers in their societies.

They were what, in more modern times (until recently), would have been called practitioners of folk medicine. They were gentle healers who worked with herbs, educated in lore accumulated over thousands of years, which is being massively reconfirmed today. They were also what we today would call energy healers, people who work with techniques that today are rapidly coming back into vogue and understanding.

In the Middle Ages the Church escalated its campaign to stamp out what they saw both as a form of heresy and as dangerous competition for the loyalty they felt the Church and pope must absolutely command. Key to their campaign was the training of Church-blessed *male* physicians to replace the dangerous female healers.

What this meant was not just a switch in gender for the Church-blessed, and thus increasingly societally blessed, main health provider. It also meant that, in contrast to the gentle, caring methods of the women healers—and despite the long proved effectiveness of their healing tradition—these Church physicians resorted mainly to two radically different healing methods.

One sanctified intervention was prayer. The idea was that if the priest or Church physician prayed for you, you lived if God so willed it; if not, too bad, you died.

The other method was composed of what were increasingly called "heroic" measures. These ranged from the exorcizing of "devils," to hacking away at, or sawing off, the offending member, hand, or leg, to bleeding with cuts and leeches. This last method took the lives of unknown numbers of victims, including George Washington, who went to bed otherwise perfectly healthy with a head cold and within days had been done in by a succession of amateur and professional bleeders.

But was this definitely the situation in Italy in the early 1600s? Records show that between 1596 and 1785, in the Venice area, there were 777 accusations of witchery, half directed toward women primarily involved with healing. "After an initial preoccupation with Protestant heretics," one of the best sources on the witch hunts tells us, "the majority of Italian inquisitional cases were directed against folk healers and diviners."

8. Was there any evidence of picnic gatherings on plateaus above the rooftops by the merchant-craftspeople class of that time? By some unbelievably remote chance could an account have survived of my rescue of the young girl who had fallen from a high-up green picnic area?

The director of the Amalfi library, Andrea Cerenza, as well as one Giuseppe Cobalto, confirmed the probability of picnics. While unable to cite a mention of picnics or other social gatherings by the merchant class in that time, they felt it was surely probable "because the Amalfi people loved these things."

Of a specific account of the rescue, of course, as I anticipated, there was nothing of this recorded in anything of any importance from that time.

9. Might I find some picture of the potentate who with grandiloquent disregard of anything but the gesture had perfunctorily handed this low person the certificate that recorded my great deed?

This was perhaps the most fascinating of my discoveries. "What does he look like, this ruler?" Nadya had asked back in Carmel. I had described him earlier as "a man in red with heavy eyebrows, debonair, with a wax-pointed mustache and fierce Van Dyke pointed beard. He looks much like the pictures of the great bass Feodor Chaliapin in his role as Mephistopheles, the devil, in Gounod's opera *Faust* . . . I have the impression he is an arbitrary figure to be feared, not liked, with a certain playful charm dancing over a cold, cold heart. He has a resemblance to a Spanish conquistador—the same faintly contemptuous look and imperious bearing shown in the pictures in history books."

I found that Italy was ruled by Spain at the time, but beyond this, even working with an assistant, I could find nothing. Then suddenly she arrived, dancing with glee—having found that in the 1600s two viceroys of Naples visited Amalfi. The first viceroy to arrive, in 1610, was Alfonso Pimental d'Herrera. The second, in 1615, was Pedro Fernandes de Castro, both obviously *Spanish* appointees of the King of Spain.

If the rescue of the young girl had taken place in 1611, Viceroy Alfonso Pimental d'Herrera, having arrived the year before, would have been the likely subruler at the time the spice merchant was honored. Moreover, I found the next best thing to an actual picture of the man. I soon found the two best-known court painters for the Spanish rulers of that time were none other than the famous Peter Paul Rubens and the very famous Diego Rodriguez de Silva y Velasquez, known generally today as just plain Velasquez.

On examining their collected works, it quickly became evident that the overwhelming majority of Spanish noblemen—and indeed every other male in any position of authority—wore the familiar combination of pointed beard and pointed mustache known today as the "Van Dyke" after the other great court painter of that time, Anthony Van Dyke. It appeared indeed to be the chief signifier of male authority or badge of office for the time. It was so prevalent in portrait after portrait that I dubbed it "the look."

The other options were to be either beardless, which was restricted to very young men, or solely wearing a mustache, which only Philip IV sported. This, I strongly suspected, was either because he had difficulty growing a sufficiently impressive beard, or because he wanted to differentiate himself in some way from his underlings and forbade the use of his style by others.

To prove this probe was conducted in the most scientific way, here is the exact count.

For *Rubens*: "the look," 4; beardless, 1; mustache only, 1.

For *Velasquez*: "the look," 11; beardless, 2; mustache only, 6.

If I ruled out men too young to be appointed viceroys to Amalfi (whose chief purpose would of course have been to look fierce enough to cow the populace), the probability was at least 95 percent that "my" viceroy, Alfonso Pimental d'Herrera, also sported "the look," which is how "I" had seen him!

10. And what of my abrupt and ironic death? Was there something brewing at that time that could inflame two young hotbloods to pull pistols and suddenly transform a peaceful picnic into the bloody murder of an elder? Could I have been caught between some raging difference between generations, classes, ideologies, or rich and poor in that time?

The answer to this question took me ever deeper into an exploration of the history of Amalfi. With an inexpressible excitement, I found the search suddenly opening into the history of the world and the dynamics of the interweaving of personality and evolution that only my book itself, *Return to Amalfi,* can begin to convey.

Had I really been there? Was this a catch point message in time for one of the restless millions of us—now ever so often haunted by a sense of being a wanderer somehow displaced and meandering like a pinball within the flashing lights, the whoops and the jolly tilting, and the warning bells of today's world?

I feel it was a journey into the Akashic field beyond all possible denial. That is, among believers. The skeptic will of course remain locked into the mind's present prison until science and society have sufficiently evolved so that thinking otherwise will be considered something akin to belief in a flat Earth.

FOUR

Running with Spotted Fawn in the Akashic Field

Stanley Krippner

Stanley Krippner is professor of psychology at Saybrook Graduate School and Research Center, San Francisco.[1] He is the recipient of the American Psychological Association's 2002 Award for Distinguished Contributions to the International Advancement of Psychology and during the same year was given the Dr. J. B. Rhine Award for Lifetime Achievement in Parapsychology. He is a former president of the International Association for the Study of Dreams, the Association for Humanistic Psychology, and the Parapsychological Association. Krippner has coauthored a number of major scientific studies and published numerous papers in professional and popular journals.

In the year 1970 I was introduced to the intertribal medicine man Rolling Thunder by Mickey Hart, one of the drummers for the Grateful Dead rock band. This introduction was to play a pivotal role in Rolling Thunder's life because, one year later, I instigated a meeting between Rolling Thunder and Irving Oyle, an osteopathic physician. Following several hours alone at Hart's ranch and recording studio in Novato, California, the two practitioners emerged arm in arm. Oyle commented, "We compared our practices. Rolling Thunder said that when a sick person comes to him, he makes a diagnosis, goes through a ritual, and gives that person some medicine that will restore health. I replied that when a patient comes to me, I make a diagnosis and go through the ritual of writing a prescription that will give the patient some medicine to restore health. In both cases a great deal of magic is involved—the type of magic called 'faith in one's doctor.'"

In 1971 I was the program director for a conference on internal self-regulation sponsored by the Menninger Foundation in Kansas. At the conference, Rolling Thunder, for the first time, addressed a group of physicians and scientists. He described the "otherworld" from which he claimed to derive much of his healing power, remarking, "Many times I don't know what medicine I'm going to use until the 'doctoring' is going on; sometimes when it's all over, I can't remember what I've used. That's because it's not me doing the 'doctoring.' It's the Great Spirit working through me."

Then, in 1974, I made my first visit to Rolling Thunder's home in Carlin, Nevada. When my friends and I arrived, Rolling Thunder was "out on a run" with the railroad for which he worked as a brakeman. However, we were warmly greeted by his wife, Spotted Fawn, who also prepared breakfast for us the following morning. Rolling Thunder, back from the railroad, joined us for breakfast, after which he invited us to make ourselves comfortable in the living room. Sitting underneath a stuffed eagle, Rolling Thunder observed, "When I find a plant I have never seen before, I can hold it in my hand and tell what its uses are. It will communicate with me. It will sing its songs and reveal its secrets." During our conversation, Spotted Fawn's ready smile, genial hospitality, and endless dispensation of refreshments kept the interchange well lubricated.

John Rolling Thunder Pope was born Cherokee but was later adopted by members of the Western Shoshone tribe. Helen Spotted Fawn Pope was a Western Shoshone. Over the years, I met several of their children: Mala Spotted Eagle, Buffalo Horse, Ozella Morning Star, and Patty Mocking Bird. Each of them eventually chose his or her individual path but remained true to the traditional wisdom taught by their parents.

Upon retiring from the railroad, Rolling Thunder organized a small spiritual community called "Meta Tantay," or "Go in Peace." Visitors, primarily from Western Europe and North America, spent various amounts of time at Meta Tantay studying Native American medicine and traditional lifestyles. It was here that Spotted Fawn assumed her own power. Regarded as the "heart" of Meta Tantay, she became the "clan mother" and had the final authority over all matters regarding women. A veto power over all community decisions is part of a clan mother's authority, but Spotted Fawn preferred to exercise a mediating role; as a result, her opinion was sought and respected on all levels.[2] Ken Cohen, a Taoist master who often taught at Meta Tantay, marveled at how "Spotted Fawn was the great organizer; without her work, neither Rolling Thunder nor Meta Tantay would have been half so effective."

SPOTTED FAWN'S ROLE AT META TANTAY

Spotted Fawn took on a maternal role with runaways and outcasts, an instructional role with women entering the "moon lodge" for the first time during menstruation, a supervisory role with the kitchen staff who prepared three savory meals each day, and a spiritual role when she told campfire stories, led songfests, and counseled troubled camp members. Some of Spotted Fawn's most memorable contributions were her spontaneous discourses on relationships between the genders, illuminated with graphic examples, practical guidelines, and dramatic case histories that were as candid as they were comical. The young women at Meta Tantay could not have had a more knowledgeable—or a more outrageous—director for their sex education classes! Dr. Jean Millay, an artist who spent a considerable amount of time with Spotted Fawn during her San Francisco visits, observed, "This lady was the prototypical Clan Mother—sweet, loving, gentle, healing, and yet imbued with a special power."

Spotted Fawn was also a ritualist, reminding her husband when a sunrise service was needed and actively participating in many of the sacred ceremonies held at Meta Tantay. She felt that rituals were one way to "spot the phonies," because the rituals the phonies performed lacked the proper context, organization, and respect for tradition that characterized authentic performances.

When it came to the preparation of food, Spotted Fawn told the kitchen crew that they were not only providing nourishment for the body but also for the spirit. In regard to "moon customs," Spotted Fawn reminded young women that this was a sacred time in which they could purify themselves and rejuvenate their strength. "When Grandmother Moon comes to visit," Spotted Fawn remarked, "it is the time for prayer and renewal." Menstruation was a time when the power of women was self-evident; to maintain the balance needed both by Rolling Thunder and Meta Tantay, during menses, women sat in the rear of the audience and maintained a low profile. They avoided sunrise ceremonies and often stayed in the "moon lodge." Spotted Fawn explained to me that this was in no way discrimination against women but recognition of their power and special needs during "moon time."

During the years of Meta Tantay, Rolling Thunder was in constant demand as a speaker; I recall introducing him to an audience of thousands in Cologne, Germany, in 1982, and to smaller audiences at Sonoma State University and Saybrook Graduate School. In the meantime, Spotted Fawn, often aided by other family members and by Meta Tantay veterans, provided the cohesion so badly needed to hold the community together during the turbulent 1970s and early 1980s. On one occasion, she traveled to San Francisco without Rolling Thunder to visit family members. I arranged an impromptu party for her; she was genuinely touched because, in her words, "No one has ever held a party in my honor before." I replied, "Well, it's about time!"

THE MOST ACTIVE RADIATION SITE

In 1955, I was working in Richmond, Virginia, and viewed a film, *The Conqueror,* a highly fictionalized version of Genghis Khan's empire-building exploits. Several months earlier a minor event in cinema history had taken place—with dire consequences for the film company. On

July 6, 1954, in St. George, Utah, the local Elks Lodge played a charity softball game against *The Conqueror* film company, which was on location. Susan Hayward kicked off her shoes, ran the bases barefoot, and scored a run. John Wayne (improbably cast as Genghis Khan, the Mongol emperor) and Dick Powell (who directed the opus) scored several runs each. Agnes Moorehead, another of the film's stars, cheered from the grandstand. Three decades later, all four were dead from cancer; half of the 200-member cast and crew had come down with cancer as well.

Fallout from the Nevada test site, called "the world's most active test site" by *Shaman's Drum* in its Summer 1992 issue, was routinely carried by winds to St. George and nearby areas. The cumulative fallout of the nuclear tests of 1951, 1952, and 1953 had covered the ground of Utah, Arizona, and Nevada in uneven blotches, most intensely in and around the area where most of the footage was shot for the movie. Many fission products like strontium 90 and cesium 137 decay slowly; they are driven below the surface of the soil by rain and snow. When the soil is stirred up, the buried poisons emerge again. During the filming, there were dusty battle scenes; during mock battles, the actors and extras rolled in the sands. Electric blowers were brought in to simulate windstorms. So much dirt collected in the costumes that actors had to be hosed down before they took them off.

The land where the tests were conducted rightfully belongs to the Western Shoshone nation, but neither the United States nor Great Britain asked permission for the tests of nuclear weapons both countries conducted there. Not far from the Nevada proving grounds, Spotted Fawn was enjoying young womanhood with members of her family. In 1984, three decades later, she was hospitalized for cancer. Once or twice each week I visited Letterman Memorial Hospital in San Francisco's Presidio where Spotted Fawn, as a Native American, was able to obtain low-cost medical care at the massive military facility. At Spotted Fawn's request, I led her through a series of guided imagery exercises to ease the pain. The most effective image seemed to be a calming, peaceful blue light. When she merged with this light in her imagination, the discomfort eased, and sometimes abated completely.

During this time, Rolling Thunder accepted Mickey Hart's offer to stay at his Novato ranch. Rolling Thunder and his associates drove from Novato to San Francisco every day to comfort Spotted Fawn, and Rolling Thunder spent his meager resources to secure outstanding practitioners from both mainstream allopathic medicine and from alternative and complementary medicine. In addition, I brought to the Presidio hospital friends of mine who taught Spotted Fawn various types of self-regulation and pain control in an attempt to reduce her discomfort. These psychologists, physicians, and psycho-physiologists had backgrounds in biofeedback, meditation, autogenic training, the relaxation response, and other self-regulation techniques. Welcoming all this attention, Spotted Fawn bloomed, and followed the instructions meticulously. Rolling Thunder complied with the suggestions as well, stating that he utilized them to mitigate the personal stress his wife's illness had induced.

On one July afternoon I spent an hour with Spotted Fawn, taking her through a series of progressive relaxation exercises, ending with the image of the blue light that had provided her so much relief in previous sessions. Spotted Fawn told me that this session had been especially compelling and that she had difficulty returning from the light. At that point, I sensed that Spotted Fawn was close to death and had made her peace with her condition. In fact, she indirectly taught me that "healing" differs from "curing." Spotted Fawn might not have been "cured" of cancer, but she had attained a profound "healing," an experience that comes from love, respect, and acceptance rather than from intellectual understanding.

DREAMING IN MEXICO

In August 1984, I was attending a parapsychology conference in Mexico. On the night of August 15, I had a dream that I had arrived at Mickey Hart's ranch. As I drove in, Rolling Thunder and his group were driving out. He had a sober expression on his face, as did the other members of his entourage. I asked, "Where is Spotted Fawn?" He turned his head slightly toward the back of the truck where I saw a wooden coffin strapped to the floor. I knew that it contained the earthly remains of my dear friend Spotted Fawn. That morning I experienced what is called a "hypnopompic image." Although most such images are visual, this one was auditory. I heard Spotted Fawn's voice telling me, in a very unpretentious way, "You know, I won't be seeing you any more."

Upon arriving back in the United States, I learned that Spotted Fawn had passed away on that very night. In retrospect, the dream and the voice were anecdotal, thus have very little evidential value for strictly scientific purposes. Nevertheless, it occurred to me that Spotted Fawn and I had been "running" in the Akashic field, and I was able to obtain information about her condition at a distance. The term "Akashic field," coined by Ervin Laszlo,[3] is very similar to what many parapsychologists such as Roll[4] and Stokes[5] and also Cheney[6] have hypothesized is a "psi field" that is trans-temporal and trans-spatial in nature.

In parapsychology—the discipline sometimes called "psychical research," or "psi research"—the word *psi* refers to anomalous interactions. Examples of these interactions are reports of such events as telepathy, clairvoyance, precognition, and psychokinesis. Remarkable healings, life after death, and past life experiences are added to the list by some investigators. Each example may be part of the "cosmic information field" discussed by Laszlo, who calls it the "Akashic field in homage to the ancient Sanskrit term signifying a space-time continuum that's all-pervasive."

Each of these experiences can be studied in several ways to determine whether it actually is trans-spatial (as in clairvoyance, the purported obtaining of information from a distance) or trans-temporal (as in bringing to consciousness events from the past, or in precognition, putative knowledge about future events). Parapsychologists use questionnaires, interviews, and field observations in their studies. In each of these cases, there is a possibility that conventional scientific explanations can account for the reported experience. Some of the possible explanations are subtle sensory or motor activity, misinterpretation, poor memory for an event, and deliberate fraud. If an investigation systematically eliminates conventional scientific explanations, that research is said to have been performed under "psi-task" conditions.

My dream on the night of Spotted Fawn's death, and the voice I heard when I awakened, were not obtained under "psi-task" conditions. However, they were meaningful and suggest that our psyches had met in the Akashic field. Laszlo's concept is one that may be able to provide an explanation for such experiences.

SPOTTED FAWN'S LEGACY

Shortly after the death of his wife, Rolling Thunder terminated much of his healing practice. He maintained an active lecture schedule, and we visited frequently, both in Nevada and in California. But Meta Tantay slowly crumbled. Without the heart of Meta Tantay to replenish the community's supply of love and energy, people drifted away, or simply spent shorter periods of time on the land. In 1997 Rolling Thunder joined Spotted Fawn in the "otherworld," leaving behind an impressive legacy that, from my perspective, was directed toward activating his clients' "inner healer," connecting it with whatever spiritual, social, biological, emotional, and mental resources were available.

In retrospect, Spotted Fawn left a legacy as well. She taught me, and many others, that it's not enough to plan and execute a project intellectually. If a project is to grow and prosper, it needs to be rooted in the spirit, in the body, in the community, as well as in the mind. And in her final months of life, Spotted Fawn bore her adversity with dignity and good humor. She was neither a whiner nor a complainer. She appreciated the help my associates and I were giving her, often joking about her condition. For example, Spotted Fawn's liver failure had resulted in ascites, the accumulation of fluid in the peritoneum, leading to an enlarged abdomen. Spotted Fawn often talked about her "pregnancy" and how people would be surprised when she gave birth! From my perspective, even in her final weeks on Earth, Spotted Fawn did give birth; she delivered a new understanding and appreciation of life to those who came into contact with her.

Spotted Fawn also imbued me with a respect for the land, for Native American traditions, and for the importance of service. Spotted Fawn was the clan mother for more people than she realized. She taught many of us about caretaking: taking care of Mother Earth, of our immediate and extended families, of our cultural traditions, and of each other.

Both Rolling Thunder and Spotted Fawn had told me about how "spiritual healing" begins with respect for the Great Spirit—the life and love that can be found in all of nature's creations. Each element of creation has its own will, its own way, and its own purpose. These ways need to be respected, not exploited, by human beings.

FIVE

My "Ordinary" Akashic Experiences

Jude Currivan

Jude Currivan is a cosmologist, sensitive, healer, author of four books (one, *CosMos,* coauthored with Ervin Laszlo), and educator. She holds a Ph.D. in archaeology researching ancient cosmologies and a masters degree in physics specializing in cosmology and quantum physics. Previously a high-powered businesswoman, her worldwide work now empowers others by reconciling leading-edge science, frontier research into consciousness, and spiritual wisdom to explain and experience the emerging awareness of integral reality.

I was four years old and snuggling down to sleep, when in the near darkness of my bedroom I felt and then saw the dim glimmer of an unknown presence. With the sense of curiosity my mother had recognized from my birth, I was more interested in this discarnate being than frightened. And that's when the guidance and teaching began.

Although I didn't feel that I "shouldn't" say anything to anyone else about the presence, I somehow knew to keep it and its guidance private. But over the following months, my parents saw an incredible flowering of my interest in things that were completely unknown to them—or indeed anyone in our family or surroundings.

Aged five, and mentored by my discarnate guide, I was already studying leading-edge science and ancient wisdom teachings. I was also experiencing what we would now term "nonlocal consciousness"—Akashic awareness beyond the limits of space and time. For me it was normal, not paranormal; natural, not supernatural.

In 1952 I had been born into a working-class family in the industrial English Midlands. I never knew my biological father whom my mother divorced when I was less than three years old. When she remarried, it was to a gentle giant of a coal miner who adopted me and my younger brother but who sadly was to die very young, when I was ten years old and my brother only seven. Our beloved mother became a single mum—highly unusual in those days. But along with facing all the material challenges of bringing us up, she always gave us an abundance of unconditional love. Her totally supportive, although I'm sure bemused, acceptance of my passionate search for universal truths was the bedrock of a quest that continues to this day.

I remember one afternoon when I was about eight years old. Mum had invited our neighbors round to hear me talk about quantum physics. I was in my element explaining the interconnections between everything within and beyond our universe. Mum had provided my captive audience with tea and biscuits. And although I'm not sure anyone understood a word I was saying, I think they enjoyed the refreshments!

My experiences of "walking between worlds" as well as my studies of ancient wisdom gave me a very early appreciation of the Akashic experience. The panoply of consciousness, the cosmic mind, which I realized I was part of, was being revealed step by step, almost as a path of initiation. During those early years I also became aware of the identity of my discarnate guide. My fascination with the wisdom teachings of the ancient Egyptians revealed him as the archetypal wisdom bringer they knew as Thoth. As I became more used to his presence I began to become aware of other transpersonal beings guiding me as well. From the beginning, this interplay of my direct access to the Akashic field and higher awareness and the validation of the knowledge gained has been the lodestone at the center of my still unfolding understanding of the irreducible wholeness of reality that, with Ervin Laszlo, we call the "whole-world."

PHYSICAL AND EMOTIONAL BLACK HOLES

As I grew up it was always clear to me that the cosmic realm of the Akashic field generated the physical world, rather than the other way around. In order to understand how, I applied to Oxford University to undertake a degree in physics, specializing in the study of quantum theory and cosmology. While I loved the search for knowledge there, the fundamental assumption of my teachers was that the material world is all there is and that consciousness arises solely as the end result of physical evolution. The limitations of their reductionist approach and its solely materialistic premise frustrated me, but they nonetheless taught me the cosmic language of mathematics and enabled me to see further into the mechanisms of physical manifestation—all crucial to my ongoing journey of discovery.

I argued that quantum theory implied and experimentation showed that consciousness was innate to the cosmos, even though I wasn't as yet ready to reveal the source of my own understanding through my Akashic experiences. But in the 1970s, as a student in the enclave of academe that's Oxford University, I had little chance of altering the entrenched views around me. I became disillusioned with the intellectual certainties of my university professors even while appreciating their own search for universal understanding. And, despite my misgivings, I had always wanted to be a cosmologist and still did.

After early success in my university examinations and winning a prestigious prize for an essay on the then emerging discovery of black holes, a combination of emotional trauma and illness ensured that my final year at Oxford ended with a disappointing degree and the closure of possibilities of an academic career. At the time—in my disillusioned frame of mind and believing that all my dreams had come to naught—I didn't yet realize that everything in life is purposeful.

Now I am able to perceive the perfection of the inner and outer journeys of discovery we call our lives. I understand that had I become a mainstream academic, my lifelong quest for a deeper understanding of the cosmos would have been limited to what has been accepted by and is acceptable to the scientific mainstream. Instead, my life journey has enabled me as a scientist to go where the evidence has led—to the emerging paradigm of integral reality.

At that time though, Thoth's guidance that had sustained and mentored me also fell away. My loneliness at this sundering of my spiritual connection continued for the next nearly twenty years of my life. I closed down both emotionally and spiritually and embarked on a path of materiality, determined to be a worldly success. Over the next twenty years I became successful as an international businesswoman. I ultimately became the finance director of music retailer HMV worldwide. Then I was appointed in early 1991 as group finance director of the Scholl foot care company, a position that made me at that time the most senior businesswoman in the UK. Scholl's operations in nearly forty countries around the world involved me not only in managing all of the financial operations of the corporation but also in the development and implementation of business strategy and the management of change.

These wonderful experiences gave me a truly global outlook and taught me how to value diversity and to appreciate and optimize everyone's contribution. What I didn't realize at the time is that this part of my life's path grounded me in a way that enabled me to integrate my subsequent extraordinary experiences into the very substance of my life, until they became ordinary, almost daily experiences.

BELIEVING IS SEEING

By my late thirties, the excitement of my materially based life was waning and I was unable to avoid the realization of my inner emptiness. Since my time at Oxford, my spiritual connection with Thoth had been distant. But that was about to change! Like a lightbulb being switched on within me, I began to "wake up" again. Step by step I rediscovered the spiritual connection of my childhood and found that my life experiences had prepared me to connect with Thoth, my other guides, and the cosmos in a much more profound way.

Over the next few years, I became spiritually ready to take a leap of faith into the unknown. At the end of 1996, having achieved all that I'd aspired to, I left the business world. I was living in a beautiful home in the sacred landscape of Avebury in southern England, where I began to connect with the land and its ancient monuments. The archangel Michael became my guide to the energies and consciousness of the living Earth and all Her realms.

In the business world I had been used to "making things happen" and focusing on results. I was continually pushing or pulling the outcomes of events and circumstances rather than focusing my highest intention and then allowing the optimum outcome to unfold. I was always "busy." My diary was filled a year ahead and I planned and managed meticulously my own work and the activities of the businesses I helped to lead. Twenty-five years of this way of being took time to release, so at first I was still trying to plan and control my rediscovered spiritual journey in much the same way.

The landscape around my home was numinous, and I started learning how to communicate with the discarnate guardians of its ancient monuments. One guardian in particular was so energetic that I felt my whole body vibrating at such a level I wasn't able to sleep. One evening after I'd been out with friends I was driving home when I suddenly heard a loud clairaudient message from the guardian telling me to go to his monument—a so-called chambered long-barrow constructed around 6,000 years ago.

The night was moonless, and in that rural setting there were no street lights. I hesitated, as I had no flashlight in the car and the monument was some way from the road. As if to calm my fears, I heard a second inner message tell me: "You'll see!" And so I did. As I made my way from the car, up the hill to the monument and into its inky blackness, I could see, literally, only one step ahead. But that one step was all I needed. As I stood alone with only the discarnate guardian for company he clairvoyantly showed me how to allow the powerful telluric energies to flow easily through my body. Afterward, as I made my way back to the car, I heard, "You only ever need to see one step ahead. Just take it." When I returned home I slept peacefully for the first time in many nights.

Such spiritual guidance, whatever form it has taken, has always held me in good stead. Indeed things have gone awry only when I've heard it but refused to "listen" and act on it. As a scientist, I've always needed to be sure that the awareness and guidance are arising from a level of consciousness and benevolence that's beyond my own ego-based perception. Over the many years since Thoth first appeared to me and with innumerable and incredible experiences, the trust accumulated has enabled me to expand my own sense of self far beyond the illusory limitations of my human persona.

Once we are open to the possibility of such nonlocal awareness, things begin to happen! The old adage of "seeing is believing" is turned literally on its head. With our acculturated veils removed from our perception, "believing is seeing." We each, however, experience "at a distance" awareness in different ways. While some become clairvoyant (seeing), others develop clairaudient (hearing) abilities, and for a few, the nonlocal sense of smell, clairalience, is experienced. But for most of us it seems that it is our clairsentient (sensing) skills that expand. Over many years my own explorations in the multidimensional realms I have experienced have combined all four, although on a regular basis it's the clairsentient aspects that have generally been emphasized.

However, a few years ago I had an experience in which clairaudience played a large part. An old friend had passed over, and on the morning of his funeral I went to the florist to choose some flowers. As I entered the shop, I heard his discarnate voice almost shout in my ear, "Rose!" As I looked around, a second message came, "Yellow rose." There were some beautiful deep yellow roses in front of me, but I sensed they were the wrong color and I gazed beyond them until I saw some pale creamy yellow roses at the rear of the shop. Sensing his approval, I "knew" they were perfect. The voice of my departed friend made it clear, though, that he only wished me to purchase a single rose. I felt embarrassed; what would his wife think of my husband and me arriving with just a single yellow rose when no doubt everyone else would be offering bouquets and wreaths! Nonetheless, the voice was insistent and I left the shop clutching the single rose.

When we arrived at the funeral, we could see our friend's wife looking lonely and sorrowful and we didn't wish to intrude. But once again his voice returned, "Give her the rose now." I walked over and handed her the beautiful rose but said, in all truth, that it wasn't from us but from her husband. Her eyes lit up but there wasn't time to say any more as the funeral was about to begin. Later we were able to speak. She told us that during their long marriage, her husband had brought her each week a bouquet of flowers comprising a dozen red roses—and one creamy yellow rose. Since his death she had felt desperately alone and had been dreading the funeral. But when I gave her the rose and spoke the words I did, she realized in that moment that her adored husband was still with her and knew that he would remain for as long as she needed him.

A GLOBAL QUEST

Some of my most significant experiences with the multidimensional realms of the Akashic field have literally taken me around the world on a global quest to understand our spiritual heritage and our cosmic destiny. They have included an abundance of synchronicities—"meaningful coincidences" as coined by Carl Jung—unexplainable within the limitations of physical space and time. Paying attention to such synchronicities and "showing up," by heeding the inner and outer paths of discovery for which they have often acted as signposts, has been and continues to be crucial in that aspect of my work that seeks to understand our human mission on Earth. In many instances they have been the portals to guidance regarding how, as part of the shift of consciousness that is gathering pace, we may heal the schisms that have dismembered our collective psyche.

My global quest began on May 4, 1998, when I went to Silbury Hill, an ancient monumental mound in the Avebury landscape, in answer to a clairaudient message of the day before. Fully visible from the summit of the hill was an enormous crop circle in the form of a glorious golden disc nearly 200 feet across in the adjacent field of oilseed rape. I'd discovered earlier that morning that it had appeared overnight. As I sat down to attune my awareness to the hill and the crop circle, I felt profoundly peaceful, as though all my inner questioning had been answered. From that simple moment synchronicities and signs began to unfold in an incredible way, guiding a six-year quest in which I traveled around the world, along with many others who journeyed with me. A series of thirteen journeys to Egypt, South Africa, China, Alaska, Peru, Australia, New Zealand, Chile, Easter Island, Hawaii, the Indian Ocean, England, and finally Jerusalem revealed our hidden heritage and became thirteen steps in our own inner transformations.

During those years, my experiences and those of my fellow travelers included profound insights and emotional healing for us personally. We also saw how the causative cocreativity of the Akashic field was manifest in the still unfolding flow and processes of reconciliation and resolution that we supported and were witness to on transpersonal, archetypal, and collective levels around the world.

I experience a fundamental aspect of the current shift of consciousness in an increasing frequency of synchronicities great and small, which provide an ongoing reminder of the all pervasive Akashic field and the cosmic mind of which we, and everything we call reality, are an integral part. One such heartfelt synchronicity recently occurred after I co-led a group of people to Egypt.

Planned for many months before, our trip had brought us on October 9, 2007, to Giza and a special access visit to the Great Pyramid. That morning, I'd turned on the television in my room to hear the news. At that precise moment, the newsreader told us that this day was John Lennon's birthday. The words of his great song *Imagine* immediately flowed into my mind, my heart surging with its simple wisdom. With just enough time before we were due to meet the group and journey on to the pyramid, I logged on to the hotel's computer and downloaded a copy of the words of the song, then literally ran with them onto the bus.

Later as we all gathered in the Kings Chamber of that mighty monument, in tribute to John and the many other peacemakers who have worked, and continue to work, for healing and reconciliation, we sang *Imagine* together, the acoustics of the chamber swelling our voices into inspiration.

A few weeks later, as I packed my bag for a trip to Vermont where I was to be the final speaker at a sacred landscape conference, I had a strong feeling that I should take the words of *Imagine* with me. After the conference organizer met me at the airport, I tentatively asked her whether it would be all right if I invited everyone to join in singing the song after my talk. She said, "Of course, we always sing *Imagine* at the end of our conferences!" At the end of the conference, I shared the story of the synchronicity as several hundred people then joined hands in an enormous circle that filled the room. As we joined together in this song of hope and peace, a wonderful feeling of joy surrounded and flowed through us.

Mainstream science would dismiss such a synchronicity—free of the apparent limitations of space and time—as mere coincidence. Indeed, the current mainstream view is that the universe is random and purposeless. But I believe this limited and limiting understanding is about to radically change. The revolutionary new vision of the universe that is emerging is revealing a fundamentally interconnected and meaningful cosmos—one that I have been privileged to explore all my life. This empowering understanding of the world fosters our innate abilities to be aware and influence our realities beyond the limitations of space and time.

This expanded understanding of *all* that we call reality is showing that meaningful coincidences are real! As we become more aware of the significance of such synchronicities in our lives, they can be signposts that help us on our spiritual path and keys to deeper insights. So when we're thinking of friends we haven't heard from for months and they suddenly call, or when a series of events magically come together, we need to take a closer look at the synchronicity. It may just be that our spiritual self is showing us how to go beyond our perceived limitations and play with reality.

A NEW WAY OF BEING

The emerging vision of the Akashic field and the integral nature of the whole-world require us to recognize that we are both the creation of the cosmos and the cocreators of our realities. In my Akashic experiences, such cocreativity encompasses many levels of awareness, from the often submerged impulses and responses of our subconscious and the "normal" waking perceptions of our human persona, through transpersonal, collective, and archetypal consciousness, to the ultimate reunion with the infinite cosmic mind.

On both personal and collective levels, our awareness is rapidly expanding beyond the limitations of the past. The shift of consciousness is offering us deeper, more empowering insights into our nature and that of the whole-world. As we open ourselves to the Akashic experience and the higher awareness that's our innate heritage and destiny, I believe we can transcend our egos—not to become "ego-less" but "ego-free." As we do so, we become ever more attuned to the unfolding flow of the cosmos and progressively more conscious cocreators.

The more attuned we become, the more our choices are both empowered and inherently altruistic. In acknowledging all that we are—both light and shadow—we transcend the judgment of polarity-based perception and begin to embody the unconditionally loving awareness of the unity that ultimately is the whole-world of the Akashic field and of cosmic mind.

This is a new way of being. Two thousand years ago Jesus said, "As I do these things, so shall ye do them and greater." And—as the elders of the indigenous peoples who still walk the Earth maintain—we are the ones we have been waiting for. The shift in consciousness is our next evolutionary leap. It is enabling us to re-member who we *really* are and to ultimately fully embody our spiritual nature and innate nonlocal connectivity within our human experiences.

Throughout my own quest and adventures, time and time again I have been reminded of the importance of melding the ordinariness of my everyday life with extraordinary exploration of the whole-world. But the deeper significance of their interweaving came to me one gorgeous morning in the midst of the Avebury landscape. As I gazed around at the beauty of the natural landscape surrounding me, my heart singing with the quiet joy of the day, I heard the words, "In the commonality of our humanity, we are all ordinary. In the commonality of our divinity we are all extraordinary."

At the time, I had been journeying around the world on the spiritual quest I would eventually relate in *The 13th Step* and also immersed in studies of ancient cosmologies for a doctorate in archaeology, so I was "fully-whelmed" (not overwhelmed but close!) by extraordinary experiences. And I was in danger of ignoring, even disdaining, my "ordinary" responsibilities and relationships. The simple yet profound words of my discarnate guide that morning shifted something within me. I suddenly became aware of what I was missing by not being fully present in *every* aspect of my life. Instead of bringing an attitude of joy, creativity, and sanctity to those parts of my life I deemed ordinary, I had been judging them as somehow less important. In that moment I appreciated that I could only embody the balance and wholeness I was seeking by perceiving the extraordinary in the ordinary and the ordinary in the extraordinary, rather like the Chinese glyph of yin and yang.

That simple insight changed my life. I began to discover ever greater happiness within myself. I wasn't *searching* for anything anymore but joyous and grateful for how each day flowered. I realized that being truly present in each moment, with whatever task was before me, enabled me to become more attuned to my higher guidance and the flow of awareness through me.

My presence in the Avebury landscape that morning was due to the archaeological field walking that formed part of my Ph.D. research. This aspect of my research was indeed very "ordinary." Day after day, sometimes with a group of helpers and sometimes by myself, I walked the plowed fields of a landscape that has been continually occupied since Neolithic times with the aim of retrieving flint artifacts to show the pattern of that early occupation. After field walking for many months and many miles, and analyzing many thousands of pieces of knapped flint, the final day of walking (thankfully) arrived. I had been helped by a number of volunteers who always seemed to pick up the interesting finds, while I had picked up nothing but enormous amounts of broken bits and pieces.

As I walked the final stretch of the final session of the final day, I asked the cosmos for something, anything! As I came to the final yard of the long, long project, literally on my final step, I looked down. There, lying in front of me, was a perfect flint arrowhead exquisitely knapped from white rock by an unknown hunter some four thousand years before. After all my hard work over so many months, I was thrilled by this small yet so extraordinarily synchronous gift.

Four years later, while coauthoring *CosMos* with Ervin Laszlo, I was writing a section on entropy and the arrow of time within our physical universe. As I struggled to find the best words to explain how the extraordinary order at the beginning of our universe enabled the arrow of time to take flight, I glanced up to where the ancient arrowhead now sits in my study and the right words too, took flight.

Over more than fifty years, my experiences and exploration of the whole-world have led me to an understanding that in every moment we cocreate our realities afresh. In every breath, heartbeat, and thought we can make new and different choices. I once wrote that

> *If then we chose fear, we can now choose love.*
> *If then we chose tears, we can now choose laughter.*
> *If then we chose death, we can now choose life.*

Now more than ever, I believe that the shift in consciousness is enabling us both individually and together to make the choices that over the next few years will determine our collective destiny. As our awareness expands, we are enabled to transcend our ego-based fears; rather than undergo a threatened global breakdown, we can instead cocreate a breakthrough. After giving away our empowered sense of self for so long to external authorities and gurus, it is now time for *us* to "guru up!"

SIX

A Journalist's Encounters with the Akashic Experience

Guido Ferrari

Guido Ferrari is a television journalist and director. He has made numerous documentaries in the arts field and has interviewed and created biographical programs on a number of well-known personalities. His biographical films and interviews include the Dalai Lama, Erich Fromm, Karl Popper, Eugene Ionesco, Ervin Laszlo, Elisabeth Kübler-Ross, and Marie-Louise Von Franz. He recently produced two DVDs on Buddhism in collaboration with Matthieu Ricard.

I am a television journalist and director, for many years associated with the Swiss Italian Television. In the course of my work I have encountered surprising experiences—experiences that confirm that we are more than mere flesh and blood, that we are linked to each other and to every single thing, experiences that testify that we can have extrasensory perceptions, which are full of meaning and provide answers to fundamental questions.

A MOST PRECIOUS GIFT

In 1982 I directed and produced a documentary on near-death experiences (NDEs). I looked for a Tibetan monk who would be able to describe the experiences of *bardo* (in Tibetan Buddhism, a state of existence between death and rebirth). I found him, and contact with him proved to be one of the most important things in my life.

I knew almost nothing about Buddhism. I remember that Lama S. suggested that I listen to some of his teachings before beginning. There were six or seven of us. In very simple fashion Lama S. spoke of love and of compassion, of poisons in the mind, of how everything changes and of the error of attachment. I remember that my heart began to open to a new warmth that filled my whole body, while my mind stayed at rest, lucid and full of space. I heard his words like an echo of a truth that I had known from time to time but that had been buried beneath negative emotions. It was a "coming home," an experience of direct union, of communication without words, of peace and harmony. I could feel that the human heart is good and perfect and that we are all one. This was an extraordinary discovery for someone like myself, raised in a rigorous scientific tradition that left little space for the heart or emotions. He had given me the most precious gift there is.

From that moment on I dedicated my life to conserving and developing this feeling of oneness. Lama S. suggested a short period of meditation. It was the first time for me. It proved to be an extraordinary experience, both confusing and mysterious, which was to help me, over the years, to understand many things. I found myself flying over mountains and then in space that became pitch black. In the sky white lines formed a square with two diagonals. I flew to the center of the white light and then crossed through a white vortex to find myself in a Tibetan Buddhist monastery, as I was to understand later. There I met old friends and schoolmates and shouted for joy as I recognized them. They shooed me to silence: a master was in a state of deep meditation, a man whose strength made a lasting impression on me.

Then there was the return trip. I saw my life a second time, I understood my mistakes, felt my anger dissolve into a healing warmth, and then I found myself in the temple where Lama S. was "singing" a prayer in a soft voice that still evokes deep feelings when I remember it. My experience in a state of meditation was similar to that of the dying. During the weeks that followed I lived in a state of open and empathetic consciousness.

I spent years trying to understand what had happened to me. It was an absolutely real experience even though I had no prior knowledge of a Tibetan monastery or of any classmates . . .

As I write these words, the memory of another Akashic experience comes back to me. This also took place during meditation with Lama S. but some time later. I found myself at the entrance to a narrow alpine valley, which had a river flowing through it, and mountains covered with pine trees. The scene changed into that of a child being born—the walls of the valley were the mother's legs and the river her vagina. The mother immediately took the child in her arms and then I saw my own mother and father as young people, not as I remembered them but as they were before I was born. Yet they were real, making love slowly and affectionately. I remember the sensation of pure love and joy that filled my heart: sensuality full of respect, experienced through a real relationship, in which there was no attachment. It was a very moving experience. I had lived pure love, the exchange, the gift. Something beautiful had been shown to me and the state of consciousness associated with it. This vision came as a big surprise: it had come quite unexpectedly.[1]

Another experience linked to this theme of birth, motherhood, and sexuality happened to me during a seminar on holotropic breathing given by two students of Stanislav Grof. I suddenly found myself in a forest, seated at the foot of a huge pole that had been painted white, in the middle of a crowd of people who were crying and praying. Our bodies were also painted white. We were taking part in a funeral rite. I looked at myself: I was an old woman whose breasts had dried up. I could feel inside me all the pain of the loss of my child, something that had happened years before. I understood that life was also the losing of what is most precious. My heart was calm; I was immersed in the most tender memories. I lived once more the exchange of affection with a young man who must have been the father of my child. This also belonged to my youth, a sweet memory from time past. I remember that within my vision I could smell the forest and could hear the funeral chants: I was there; it was real.

Some time later I was listening to the radio when I heard a melody sung by a tribe of Pygmies and I burst into tears. Despite the reality of the episode, there was no evident historical connection to it. It felt like an old memory or one from a parallel life.

MY WAY THROUGH SOUND TO HIGHER STATES OF CONSCIOUSNESS

The next Akashic experience involved the world of sound and testified to the importance of Robert Monroe's discoveries. Monroe had created sounds—the so-called Hemi-sync sounds—that facilitate the synchronization of the cerebral hemispheres. They contain frequencies corresponding to different states of consciousness. It is possible to be attuned to this sound and thus gain access to higher states of consciousness. (This is the electronic era's version of the shaman's drum, the mantra, the singing of psalms, and Gregorian and Sufi chant.) I went to see Monroe in Virginia, where I interviewed him at length and we had long conversations. For the first time after meeting Lama S., I had found someone who understood my experiences and was willing to talk about them.

Monroe was a down-to-earth engineer and researcher, a pragmatic individual not prone to speculation. He had had a number of out-of-body experiences (OBEs) that led him to reflect on the nature of reality and of consciousness. He had built a laboratory in a cabin where out-of-body experiences are facilitated through sound. Thanks to his own experiences and those of a number of "explorers," he realized that it is possible to move in full awareness in this and other dimensions, outside of the limits of time and space.[2]

During my stay at the Monroe Institute, I had further Akashic experiences. Exposure to Hemi-sync sounds opened up the world of interior experience, leading me and others around me to higher states of consciousness. The process begins in a state of deep relaxation in which the body sleeps and the mind stays awake; then comes an expansion of consciousness that goes deeper and deeper until further dimensions are entered, beyond space and time.

The experience that made the most impact on me was one that brought back, with full emotional impact, a critical moment from my childhood. I was three or four years old when—following a bout with diphtheria at a time when there was no penicillin or antibiotics—I was taken to the hospital in a critical condition. I remember my mother's helplessness, which I took to be her abandonment of me. I remember meeting my grandparents, who were already dead at the time, in a garden and how they told me to go back to my bed at the hospital. Later I came to understand that I had had a near-death experience.

At the Monroe Institute my experience began while I was getting ready in the morning; it started with a feeling of being touched by my mother, touched like she had lovingly touched me when I was a child (my mother had died a few years previous to this). The feeling continued at breakfast when I remembered how I had refused to eat when I came out of the hospital, rebelling against my mother. The trauma of feeling myself abandoned and the experience of being near to death, with a lowered body temperature, had created the sensation that the warmth of life was outside me, that it was all inside my mother who had taken possession of it.

Then, during a session at the Institute, in which I ventured into space, I had the sense of being far away, in a night full of stars in which beautiful galaxies could be seen. I felt I was enormously big, holding the whole universe in myself: the stars, the galaxies, the sun . . . everything was inside me. I had the distinct sensation that the earth was my heart and the heat of the sun was within it. The warmth of life had returned to me physically and emotionally. It was within me, and I understood that I could take care of it, make it grow, and give it to others.

In a subsequent session I asked what I could do to increase the warmth of the heart. The answer I received was, "Forgive," because "the universe responds." In fact, on a closer look all the experiences I have described here are responses. There are times when real, tangible messages arrive that I can hear inside myself, and sometimes the messages come in the most surprising fashion, such as when an Indian mystic appeared; he was engaged in cooking and told me not to disturb him with questions to which I already knew the answers.

These and other Akashic experiences during my life as a TV journalist and director have taught me a great deal. They have lent unsuspected meaning to my life.

Part Two

WORKING WITH THE EXPERIENCE

SEVEN

The Living Classroom

Christopher Bache

Christopher Bache has been a professor of religious studies at Youngstown State University for thirty years and more recently adjunct professor at the California Institute of Integral Studies. He was director of transformative learning at the Institute of Noetic Sciences. His teaching focuses on Eastern religions, psychology of religion, and transpersonal psychology. Bache is the author of three books.

It was an ordinary day in the classroom, much like other days at the university in Ohio where I teach in the department of philosophy and religious studies. The lecture was finished and the room was emptying when a student came up to me and said, "You know, it's funny that you used the example you did in class today, because that's exactly what happened to me this week." Then he described his recent experience and it was, indeed, a perfect match.

I had been searching for an example to illustrate a concept I was trying to convey to the students. Quickly running through the possibilities in my mind, my stream of consciousness had paused, and out of the stillness an example I had never used before rose from some deeper place. "Try this . . ." it said. I used it, and it worked. The students seemed to get the point, and the lecture continued. But what had been a randomly chosen example for other students in the room cut close to the bone for this particular student. He heard his own recent life experience coming back to him in my words, and it grabbed his attention. It was as if he had been extended a personal invitation to get more deeply involved in the course, and he did.

The first time this happened twenty-five years ago, I brushed it aside as mere coincidence, as all good academics are trained to do. We are taught that our minds are fundamentally separate and discrete entities, one mind per brain. Any suspected overlap or bleed-through between minds is said to be impossible—an illusion, a fiction of circumstance. But it happened again a few months later, and again after that. In the following years, it became a not uncommon occurrence in my classes, happening just often enough that I couldn't dismiss it any more. My students were finding bits and pieces of their recent life experience, or the experience of a close family member, in my lectures. Without my intending it, without my even being aware when it was happening, my consciousness seemed to be tapping in to some kind of informational field that held their life experience. But how and why was this happening? Both as a philosopher interested in consciousness and an educator, I had to understand what was driving these events.

This was the beginning of a long odyssey with my students, a journey of discovery that lasted decades. Eventually it led me to a new understanding of the dynamics of collective fields of consciousness and to recognize a true *collective intelligence* operating in the classroom.[1]

It goes without saying that most of my university colleagues would have advised me not to pursue this line of inquiry. Being the good academics that they are, they know that the reigning materialist paradigm tells us that these are just random coincidences. When you think of all the life experience tucked inside our students, surely we're going to bump into someone's experience some of the time. If you lecture to a hundred people week after week, sooner or later you're bound to hit a few bull's-eyes even with your eyes closed. Do the math and it looks less significant than it feels. But the fit between life and lecture was so precise and repeated itself so often in my classes that eventually I was convinced that something more than chance was operating here.

RESONANCE IN THE CLASSROOM

One evening I was teaching a night course on Eastern religions. There were about thirty students in the class. In the middle of my lecture, I found myself taking a little unexpected detour in which I told a story about a Zen master who had an accurate precognitive perception of his impending death, similar to the accounts Sushila Blackman later reported in her book *Graceful Exits*.[2] I had never talked about this subject in this particular lecture. It was just a little aside, something dropped in to add some anecdotal interest to the discussion we were having about the capacities of the mind from an Eastern perspective.

After class, an elderly, silver-haired lady came up to talk with me. I had never seen her in class before. She was not part of our course but had been brought by her friend, another elderly woman who was taking the course. Her friend had dragged her to class that night because she was worried about her. Her husband had died three months before, and her friend thought she was languishing at home and needed to get out of the house.

In the conversation that followed, she told me this story. Her husband had been a used car salesman in good health. Shortly before his unexpected death, he had cleared out most of the cars from his lot without any explanation and gotten all his financial papers in order. A few days later, he and his wife were watching TV when he put down his newspaper and, in a way that was quite out of character for him, turned to his wife and said, "Darling, I just want you to know that if I died tomorrow, you've made my entire life worthwhile." A week later, he died in his sleep.

What she wanted to know was whether I thought her husband might have been unconsciously aware that he was going to die, like the person I had described in my lecture. I said that it sounded like a possibility, and this thought was comforting to her. This led to a longer conversation in which she described the challenges and opportunities his passing had created for her. In the midst of her grief she also realized how overprotective he had been and how she was now being given the opportunity to develop herself in ways that his well-intentioned care had always prevented. In the middle of the conversation, she suddenly decided to come back to college. She did so and thrived here for several years.

When these things first started happening in my courses, I was in the early stages of my career. I was also in the early stages of my spiritual practice, and here the plot thickens. To make a very long story short, as my spiritual practice deepened through the years, these synchronicities became more frequent in my classroom. It was as though by entering more deeply into conscious communion with the deeper fabric of life, the sinews of that fabric were being activated around me. Not only were these cognitive resonances occurring more frequently, they also began to touch sensitive areas in my students' lives, as the above example indicates.

It was as though a radar had been activated that was operating below the threshold of my awareness, a radar that zeroed in on some part of their life that was hurting or constricted. Sometimes it touched a question they had been holding for years or triggered an insight they had been searching for, something they needed to find before they could take the next step in their lives. Sometimes it lanced a private pain that had been festering inside them. In this mysterious communion that opened between my students and me, it was as if their souls were slipping messages to me, giving me hints on how I might reach them—telling me where they were hiding, where they were hurting, and, most importantly, what ideas they needed in order to take the next step in their development.

As the years passed and the process deepened, my students also began to have unusually deep experiences around some of the concepts I was presenting in class. It was as though their lives were being activated by more than just the ideas themselves, as though they were somehow being touched by the actual *experience* of these realities that now lived in me to some degree because of my practice.

As a professor of religious studies, I have taught more sections of "Introduction to World Religions" than I care to remember; it's a staple in our department. My approach to this course has been deeply influenced by

Huston Smith and the scholarship surrounding the perennial tradition—ideas so essential they keep showing up in multiple cultures throughout history, like perennial flowers that reemerge each spring. When students hear the perennial truths of the world's spiritual traditions simply spoken, when they are reminded of things long ago forgotten but always present at the edge of their awareness, there is sometimes a spark of recognition that can explode into a flame. This flame is contagious and sometimes stimulates sympathetic resonances with other students in the room. Students may collectively feel their energy shift to higher centers of awareness, though they may not understand what is happening at the time. Symptoms of chakra-opening and kundalini-type arousal may begin to manifest. Energy runs, hearts open, and insights arise.

These can be very powerful experiences for students. One sophomore described such an experience in an essay she wrote at the end of a course. It happened to her on a day when I was describing the Buddhist understanding of the relationship of individual mind and nondual consciousness. To convey this point, I sometimes develop the analogy of a tree, contrasting leaf-consciousness (the personal mind) with tree-consciousness (nondual consciousness). In this exercise, I ask the students to imagine that the leaves are individually aware but not yet conscious of the life of the tree they are part of, until the moment of breakthrough. It's a powerful exercise that I save until I think the class is ready to absorb its full import. On this particular day, this young woman experienced the following:

> The thing that hit me the hardest of all that we talked about in class was tree-consciousness and leaf-consciousness. It was what brought everything together for me. What made me understand everyone's interdependence and stopped me from living in fear. I was so moved that it took everything that I had not to cry in class, not from sadness but from being hit by a life-altering realization. It made an emotion rise in me that I had never felt before and I wasn't really sure how to react to it.

Another student, a woman in her midthirties, summarized a similar experience in a different class in the following way:

> Sitting in class, I felt like I was inside one of those glass ball snow scenes that folks use as paperweights. Shake the ball and mass confusion begins with flakes of fake snow swirling all around. . . . I couldn't hear the lecture. My mind struggled to focus and stay with your words, but I was missing it.
>
> Later . . . at home . . . alone. It would all return to me, the lecture. Mostly feelings. Tears. Recognition. Understanding—after I let it simmer for a while. Realization that if I didn't grab at it, it would be there waiting, this knowledge, these tiny bright spots of revelatory insight. I'd journal. I'd cry. Sometimes light and gentle, warm feel-good crying. Sometimes sobs, wracking and exhausting. I THOUGHT I WAS LOSING MY MIND A FEW TIMES.
>
> Instead of hearing your lectures with my brain-mind-intellect, I actually heard you from somewhere else. Heart-soul maybe? Ears of a type that I hadn't been exercising. They had atrophied. You gave them a workout. Or the class field was so intense that it penetrated my controlling dominant brain-mind and vibrated my heart-soul like cardiac shock paddles to bring it to life.
>
> The result? I'm becoming who I was long ago. The field bypassed my intellect and went directly to my heart to pry it open. . . . I now know what I had deeply buried in me for years, and the gift of the pick and shovel for the ongoing process comes from being in the energy of the folks in our classroom. It didn't come from me alone.

I should mention that it wasn't my intention to trigger such deep existential reactions among my students. In fact, fearing that they were out of place in a university setting, I often tried to damp them down. But I found that this was impossible without damaging the teaching process itself. Whenever my students and I would gather and simply cover the assigned material, these things would spontaneously occur without my consciously intending them. It was as though fire was lighting fire.

When we would simply focus on the task of sharing understanding, these *resonances of living experience* would spring up unpredictably—not always but often—drawing the students into heightened states of awareness.

As an academic, I was so conditioned by the atomistic, Newtonian-Cartesian paradigm that it took years before I was able to admit what now looks to me to be the obvious and natural interpretation of these events—that beneath the appearance of separation, our lives are deeply intertwined with those around us, and that my spiritual practice outside of class was somehow triggering these incidents of resonance in my classroom. This was happening not through my conscious direction but involuntarily, through some form of energetic resonance. The transpersonal states of consciousness opening in my practice at home seemed to be activating the meridians of a latent field of consciousness that included my students.

These synchronistic connections became particularly pronounced during a period of several years when I was undergoing a series of powerful transformative experiences in my practice that were breaking me down at very deep levels. The details of these experiences are not important and are described in my book *Dark Night, Early Dawn.*[3] All spiritual traditions describe a phase of inner work that involves dissolving the boundary between self and other. They describe a membrane that marks the boundary between one's individual mind-field and surrounding fields of awareness. On the near side of this membrane, the world appears to be composed of separate beings, each with their seemingly private existence. On the far side of the membrane, the world appears as an integrated whole, a continuum of energy that's unfathomably complex and extravagantly beautiful. Hence, this boundary membrane is often described as the domain of death and rebirth, death to the prison of the private self and rebirth into a larger order of wholeness that underlies and saturates life's diversity. When a practitioner is transitioning through this territory, standing at the interface of these two paradoxically compatible realities, powerful synchronicities with surrounding persons sometimes manifest.

As my inner work came to focus on this boundary, some of my students began to simultaneously undergo particularly difficult challenges in their lives. Most of my students didn't enter these waters, of course, and passed through my courses untouched by these dynamics, but some did enter them. Those who did so sometimes felt themselves coming to a breaking point in their lives or a moment of supreme risk-taking. It was as though they and I were being drawn together through a collective death-rebirth vortex, a vortex that was breaking all of us down in different ways, uprooting deeply buried pains, and crushing restrictive barriers in our lives.

Drawn into deep personal transformation, some students chose to end bad marriages or to heal wounded ones. Others left careers they had outgrown but were still holding on to. Some began to confront their addictions and others to re-approach persons from whom they had been estranged for a long time. One woman in her midforties hints at the profound disruption of her inner and outer world that occurred during this period when she began to spontaneously recover painful memories of child abuse, in a course on Buddhism of all places:

> During and after having been in your classes, my internal world became increasingly chaotic as demons from painful psychological gestalts began to emerge, and eventually colored my external world too, challenging everything I thought I was and dissolving familiar reference points. . . . As I struggled to break through powerful gestalts of pain, you spoke to and nourished my soul, making it possible for me to move more deeply into my spiritual journey.

While these kinds of responses might be expected in certain types of courses, such as a counseling course, this was not the case for the kind of courses I was teaching. Rather, these events seemed to be the *indirect effect* of our simply coming together to study. It was not the content of the course that seemed to be driving these effects but something deeper. I believe it had something to do with the juxtaposition and interaction of our life energy at some fundamental level.

Whatever description physicists eventually give this fundamental level, these experiences convinced me of a simple fact. *Clarified states of consciousness are contagious.* My efforts to realize deeper states of being seem to have caused my person to begin acting as a kind of lightning rod triggering sparks of a similar awakening among those students who were receptive to this influence. This is an utterly natural phenomenon and unstoppable effect. Our spiritual ecology simply does not permit private awakening.

LEARNING FIELDS AND GROUP MINDS

In addition to the experiences already mentioned, there were other anomalies taking place in my classroom that pushed me to look beyond the student-teacher relationship and to consider the collective dynamics of the *class as a whole.*

Perhaps the most important observation that pushed me toward a collective reading of these events was the sheer magnitude of the forces that seemed to be involved. Too many people's lives were being too deeply affected for me to conceptualize what was taking place solely in terms of resonances with my individual energy. If my person was in some way a catalyst for these experiences to surface among my students, what was actually surfacing was something larger than I could be generating. As I made the shift to thinking of this larger "something" in terms of a collective field of consciousness, a variety of conceptual and experiential pieces began to fall into place.

Students were becoming more porous not only to me but to each other. They sometimes showed up in each other's dreams in significant ways. Synchronicities between them were increasing, and life-expanding coincidences were becoming increasingly common in my courses. As one male student who returned to college after a twenty-year absence reported to me:

> Each quarter seemed to bring new and unexpected changes and synchronicities. I entered into a web of personal relationships and

> meetings with people that profoundly influenced my life. I was "finding" individuals whose circumstances were eerily similar to my own; people who knew friends of mine from obscure places in the world; people who seemed to be reading the same books at the same times and having experiences that were transforming them in the same shattering yet exhilarating ways.

A female student sent me the following description of the connectivity she experienced with other students during the same period:

> All of us who have been in your classes feel a deep connection to one another. We don't know what it is. We only know that it is there. All that I know is that I have felt something binding us all together. I remember things going on around me in class with the other students. We were sensitive to each other's thoughts and feelings. . . . I always wondered if you knew what was going on because you never said a thing in class!

Sometimes insights surfaced in the room that seemed to come not from me nor from any individual student but from the strength of our combined *collective awareness.*

This is a very subtle but distinctive experience. Sometimes when I am simply doing my job covering the day's assignment, it's as if the floor suddenly falls away. The atmosphere in the room becomes supercharged, and everyone seems to congeal into a unified state. My mind becomes unusually spacious and clear, and my students' eyes tell me that they have moved into a particularly receptive state. Our hearts seem to merge, and from this open field of compassion comes a slow stream of thoughts that I, as spokesperson for the group, unfold and work with.

In these transient moments of heightened awareness, I sometimes have the acute sensation that there is only one mind present in the room. It's as if the walls that usually separate us have become gossamer curtains. Individual persons melt into a softly glowing field of energy, and this unified energy thinks and feels and hungers to speak. Because this field incorporates the life experience of everyone present, of course we sometimes find the details of our separate histories surfacing spontaneously in it. Because it embodies our private hopes and fears, of course we are sometimes deeply touched by what comes out of it.

And then there was the strange way that my students began to periodically "jump forward" in their learning en masse. I found that periodically I had to adjust my course material because the students seemed to have taken a quantum step forward in their receptivity to the ideas being presented. After years of using a carefully designed roadmap to achieve a specific intellectual outcome, a new crop of students would signal me that they were already weeks ahead of me in the syllabus. It was as if they had found a shortcut to certain conclusions and no longer needed to be taken the long way around. Of course, many factors might have contributed to this development, including general shifts in cultural insight, a self-selecting student population, and simply improved pedagogical delivery. But after watching this cycle repeat itself numerous times through the years, I became convinced that more was going on than just this. These shifts took place too suddenly and too frequently for them to be fully accounted for by these explanations.

Slowly I began to recognize that there was a *meta-learning* taking place behind the scenes, a pattern of learning that ran deeper than the learning of individual students. I began to recognize that there were *learning fields* growing around my courses, fields of consciousness that registered the learning taking place semester after semester, fields of influence that were making it easier for subsequent generations of students to learn the same material.

Readers familiar with Rupert Sheldrake's pioneering work on morphic fields will recognize his influence here.[4] Sheldrake helped me make sense of these phenomena by allowing me to see them as symptoms of an emerging collective mind. His work gave me permission to take the radical step of recognizing that not just individual persons with complex nervous systems have minds, *groups also have minds.* My experience working with students year after year pushed me to take his revolutionary hypothesis even one step further. Even *transient groups* can manifest a kind of group consciousness under certain conditions. These conditions are: (1) collective intention focused in an emotionally engaging group project, (2) a project of sustained duration, and (3) repetition of the project in approximately the same form many times.

Recognizing the fieldlike nature of mind opens the door to a new generation of insights into the collective dynamics of consciousness and a new set of pedagogical strategies for teachers, and anyone else who works with groups. Because consciousness is a field and fields are by nature porous, the transformation of the individual cannot be isolated from the transformation of others nearby. As one opens to the *depth* of consciousness, one simultaneously activates the innate *breadth* of consciousness.

Recognizing the fieldlike nature of mind also makes it easier to understand how minds can link together, even "phase lock" with other minds to form larger operational wholes. This tendency to sync with nearby systems is not unique to consciousness but is characteristic of nature as a whole, and probably underwritten by the space- and time-transcending Akashic field itself.[5] As Steven Strogatz says in his book, *Sync,* "For reasons we don't yet understand, the tendency to synchronize is one of the most pervasive drives in the universe, extending from atoms to animals, from people to planets."[6]

Lastly, these insights into the collective dynamics of consciousness don't cancel or negate the individuality that we prize so highly in the West. A growing body of evidence suggests that what we do affects others and what others do affects us, and yet within this matrix, individuality is not suffocated but paradoxically liberated into deeper forms of self-expression. The self that sees and consciously participates in the porosity of being becomes more than the self who didn't see it. Experientially opening to the Akashic field that surrounds us melts the boundaries of the private ego, bringing about the "death of self" spoken of so often in the spiritual literature; but as the ego dies, a new form of individuality is being birthed, not an isolated individuality but one that thrives in this deeper give and take. In the end, I think we will see that expanding the breadth of our conscious participation in the surrounding matrix of life brings forward a new depth of personal presence in history.

EIGHT

Healing Over Space and Time

Maria Sági

Maria Sági holds a Ph.D. in psychology from the Oetvös Lóránd University of Budapest and is an associate member ("candidate") of the Hungarian Academy of Sciences. She was scientific collaborator of the academy's Institute of Sociology and serves as scientific director of the Club of Budapest Foundation in Budapest. She is the founder and director of the Körbler Institute Hungary, with an active practice in New Homeopathy and information medicine.

It's been twenty-five years now since I discovered that I can heal in ways other than through the methods of conventional medicine. This discovery was a surprise: as a child I hoped to be a pianist, and later I studied to become a psychologist.

I was a fifteen-year-old piano student when I suffered a nerve injury in gym class and was unable to play the piano for over a year. I couldn't write, lift a weight, or do housework with my right arm. This decided the rest of my career. It was difficult to accept at first that I had to alter my career plans, but I turned to psychology and enrolled in the university. My goal became to understand the motivations underlying human behavior.

At the time, for their first four years of study, psychology students took courses in medicine together with medical students. I enjoyed studying these subjects immensely, as I believed that whoever intends to research the mind should be familiar also with the physiology of the body.

During my fifth year in university my passion for music resurfaced. I began to investigate the effects of music on the mind, working with college students, musicians, painters, and others, trying out various methods, such as projective tests, association analysis, and music painting. For the next ten years I did research in music psychology, art psychology, and social psychology and received a doctoral degree at the ELTE University of Budapest. Subsequently I received a second doctoral degree qualifying me as associate member ("candidate") of the Hungarian Academy of Sciences. My research interests marked the rest of my life.

MY FIRST ENCOUNTER WITH ALTERNATIVE FORMS OF HEALING

From the time of my adolescent years and for many years afterward, I had minor but vexing problems digesting the food in the canteen of my school and later in my workplace. It felt like I had a stone in my stomach for hours, followed by a sudden pain and the feeling of acid scorching my stomach. This happened every time I ate at these places. A medical checkup didn't show anything wrong, and tablets designed to neutralize excess stomach acid provided only temporary relief. I tried the medicines prescribed by the doctors, but they didn't help either. I then consulted a wise old priest who was said to be able to heal with a rather special method. "Father Lajos," as he was known, used medical dowsing to diagnose the problems of those who consulted him. He selected a particular diet and a combination of plant-based remedies for his patients with the help of a pendulum. His therapy worked.

My life changed dramatically when I began to follow his method. I adopted his prescribed diet, which avoids meat, milk, bread, and sugar.

Alongside my scientific research, I studied phytotherapy and medical dowsing, and then I discovered macrobiotics. Following my study of macrobiotics at the Kushi Institute of Amsterdam, I set out on a two-fold career path: scientific research and alternative healing. I prescribed a combination of diets and phytotherapy for those who consulted me. After a while, following the first consultation, I was able to treat patients who lived at a distance without having to see them again in person, using the pendulum, as I learned from Father Lajos. It gave me great joy to see them recover quite quickly. I devoted more and more time to healing.

I was fortunate to meet the Austrian technician and innovative healer Erich Körbler in Vienna. He developed a healing method called New Homeopathy. Körbler diagnosed the condition of his patients according to the principles of Chinese medicine, using a specially designed dowsing rod that oscillates and indicates the condition of the patient. This enabled Körbler to obtain a precise and detailed picture of the energy-state of the patient's body, which indicated how this state deviated from the normal state of health.

Erich Körbler died in 1994, but I, together with my brother István, have continued to teach, apply, and develop his method. I have been giving seminars on the Körbler method and on information medicine, its further development, in Germany, Switzerland, and Austria. ("Information" in "information medicine" refers to the way both diagnosis and therapy are carried out: not by ordinary physical and biochemical means, but by reading, and affecting, the information that is intrinsic to the body.) By using information for diagnosis and treatment, I could practice my healing method also from a distance. I would examine the seminar participants who wished to consult me first in person, and then continue to treat them from a distance.

In the mid-1990s I studied classical homeopathy and became acquainted with the Psionic Medical Society of England. This society, made up of reputed and accredited medical doctors, worked exclusively through remote healing, combining Western medicine and classical homeopathy with medical dowsing. I collaborated with the society for nine years. The combination of the Psionic method with Körbler's method has been the basis of my healing work ever since. It has opened fresh possibilities for me and provided some mind-boggling perspectives.

CASE HISTORIES FROM MY EXPERIENCE WITH SPACE- AND TIME-TRANSCENDING HEALING

I have been practicing the kind of healing that transcends the limits of space and time for many years, and during this time I have had many remarkable experiences. I cite a few cases here.

For many years now, I have been waking up in the morning or at the break of dawn with the next step in healing a patient coming into my consciousness. The name of the patient appears, together with the remedy or treatment he or she needs. If a homeopathic treatment is indicated, the appropriate remedy comes to my mind, usually with the required potency. I can usually identify the correct remedy for a patient even if he or she hasn't consulted me directly but has instead consulted a healer who then called me for advice, giving me the name of the patient and the nature of the problem.

There are other variations on this faculty. Erzsébet—who has studied in Tibet and is a master of healing through the use of traditional Tibetan symbols (a discipline that I have studied with her)—would often call me on the telephone. (I should add that these symbols proved over the centuries to have a subtle but remarkably constant effect on the functioning of the body. They appear to interact with the information in the biofield that governs the bodily processes.) She called one morning as I was getting ready to leave. She told me which Tibetan healing symbol I should apply that same day. All day I was wondering why she would have said this to me. In the evening I took my dowsing rod and began searching for an answer. A close relative, who lives 200 kilometers from me, appeared in my mind. He was not well and asked for help. The symbol Erzsébet gave me was this relative's remedy.

The next morning I thanked Erzsébet for her guidance. At this time she was in a hurry, but nevertheless she brought another symbol to my attention. That night I searched for the person who would need this particular symbol and realized that it was the same person. Both the first and the second symbols proved of vital help, as my relative was in a rather poor condition at the time. One evening last month an old friend of mine called around nine o'clock, reporting a strong pain in his gums next to the last tooth. He had already tried applying an antiseptic gel but didn't obtain relief. I examined him through the remote dowsing method and sent healing information with a specific validity of seven hours. On the following afternoon my friend reported that at the beginning of the night he had still been in pain, but a little while later the pain subsided and he could sleep. After breakfast, the pain returned, probably due to eating dry toast. I then sent a modified form of healing information and asked him to report back to me if there was a change. We spoke on the phone again next day, since we had other matters to discuss. When we had finished, I asked about the pain in his gums. "Oh, I almost forgot about it," he replied. "Last night the pain went away completely."

Another case was that of Veronica, a thirty-two-year-old dentist and longtime friend. She asked her father to call me as she was in Geneva at a professional meeting and had developed red and swollen eyes and burning abscesses. She looked as if she had cried all night and felt that she could not be seen in that condition. I examined her through remote dowsing and sent her healing information, and I also told the father to tell her how to prepare a remedy on her own. The remedy—consisting of information coded in a glass of water—was to be prepared twice within a five-hour period, and on each occasion it was to be used in conjunction with different healing information. The following morning her father called again, reporting that his daughter's eyes had still been itching during the evening, but she had woken up in the morning completely recovered. In this case, although the complaint was inflammation of the eyes, I didn't treat the eyes but instead strengthened her immune system. The inflammation was a symptom of the irregular functioning of her small and large intestines, and when her immune system was strengthened, it overcame the problem on its own.

An interesting case is that of a boy born with a birth defect. Balázs was brought to me at the age of two with a severe problem of metabolism (hyperammoneamia). His condition hadn't responded to conventional treatment, and the family doctor recommended that the boy be admitted to an institution for sick children. The parents didn't accept this advice and instead brought the boy to me. His metabolic problem responded to a macrobiotic diet, homeopathic remedies, and healing information.

However, Balázs also had periodic anger fits during which he knocked his head against the wall and gnawed his knuckles and fingers until they were bloody. These crises would last about half an hour. The mother would call me and we would address the problem through remote treatment. The boy would calm down, even though he was living with his family in a town 150 km away. Over the span of several years this treatment began to produce notable results. Balázs started to speak when he was six years old. His first sentence was: "The clouds are floating beautifully in the sky." Today he is sixteen and attends a school for retarded children. Not long ago he came in third in a horse race intended for fourteen- to sixteen-year-old healthy children.

TWO CASES WITH A PECULIAR TWIST

In the following instances the specific remedy that turned out to be the right one came into my mind spontaneously, even though I was not specifically testing for it.

One of these cases concerned Balázs' mother. She developed emotional problems owing to the constant crises with her son and called me one day to ask for help. At the time I was engaged in moving, so I couldn't devote sufficient time to her. I directed her to a homeopathic physician who worked with a bioresonance machine. I was in the shower the night before she went to consult him when the name of a homeopathic remedy popped into my mind: *Staphysagria*. I knew at once that this was her remedy and that she should take it in the M (1,000) potency. I thought of telling her that she didn't need to consult the physician, but decided not to interfere. The following evening, after she had her consultation, she called and told me that the bioresonance machine came up with a remedy that the doctor had then prescribed: it was *Staphysagria* in the M potency.

The next case concerns Sándor, who was fifty-nine years old when he became my patient nine years ago. For the previous twenty years we had been neighbors in a small town in the mountains of Mátra, where he lives with his family and we have a vacation cottage. Sándor grew fruit as a hobby in his hillside garden. He heard of my work with the dowsing rod and asked me to test what minerals he should give his fruit trees. He mentioned that for the past eighteen years he had been suffering from a chronic inflammation of the joints. The pain would flare up unexpectedly, in different joints. We set up a time when I would test his trees, and regarding his problem, I suggested that he should let me know when the pain flared up again.

A few weeks later I accompanied Sándor to inspect his fruit trees. That morning he had begun to feel the pain in his arm; his wrist slowly stiffened so that even driving two kilometers to his garden proved difficult. The swelling was not visible on his wrist, so he didn't talk about it. But as we were inspecting the trees, Sándor suddenly yelled and pointed to his left wrist. There a red lump began to form and within ten minutes became as large as a walnut. As I looked at this remarkable phenomenon the words *Phosphorus M* came into my mind. I felt convinced that this was the right remedy. As I didn't have it with me, I started to send the corresponding energy into the hard, red lump on Sándor's wrist. I remember feeling completely drained as the energy passed through me.

The next day I had to leave town. When we met in the autumn Sándor told me that to his great delight he was completely healed. He had no complaints until about three years later, when he overexerted himself by shoveling while helping to build his house. His wrist became painful again. He turned to me immediately for advice. This time, however, the pain responded to healing information and to the rest I had advised.

In view of his remarkable healing, I decided to construct a detailed case history. Sándor was about forty years old when suddenly all his joints, except those in his spine, became inflamed. The pain in his fingers, wrists, elbows, shoulders, toes, ankles, knees, and waist became almost unbearable. His feet became so weak he could hardly walk. In addition sometimes a hard, red lump would rapidly form on his wrists and cause immense pain. It would not respond dependably to medical treatment—sometimes the swelling went down, and sometimes it didn't.

After a year and a half, Sándor's doctor decided to stop treatment by conventional medicines and gave him *Auredan* (gold) injections once a week. He advised Sándor to quit smoking, and he did. Sándor then had hydrotherapeutic treatment for three weeks. His complaints subsided. The acute state of his illness was followed by a chronic state, in which only one joint would become inflamed every ten days or so, quite randomly. Because the inflammation only affected one joint at a time, he was capable of working. It was impossible to say what caused the inflammation. It could have come from changes in the weather, or from having a simple cold. It appeared also to occur without an obvious reason, for example, after eating chicken. This condition had persisted for about sixteen years. I asked him what had preceded it.

Sándor said that when the symptoms first appeared he had just eaten chicken in a restaurant. The chicken leg was well done on the outside, but closer to the bone it had a distinctly bad smell. He had stopped eating it and so had not gotten sick. A light went on for me suddenly. The reason *Phosphorus M* was so successful was because it's a detoxifying agent. Taking it helped his immune system to balance the negative information he received with the rotten meat. When I had thought of this remedy I had no knowledge of the problem of the rotten chicken meat. Yet, thanks to this sudden and spontaneous intuition, Sándor's eighteen years of suffering came to an end and the pain hasn't recurred for the past nine years.

In the case of chronic diseases, it's usually necessary to trace the original cause of the illness to effect a healing, even if the cause dates from the distant past. The next two case histories exemplify this.

János, a man in his seventies, had suffered from neurodermatosis for twenty years. According to a prior diagnosis by his doctors, he was allergic to some seventy-two different foods. I didn't try to cure his allergies by working on his symptoms but instead sought the underlying causes. I found that János had suffered a trauma when he was five weeks old and that this trauma was connected with his present illness. I led him back to this early stage of his life and sent the indicated healing information. During the time I sent the information he cried loudly and waved his arms like an infant. Afterward he calmed down and became still. He accepted the diet I suggested as therapy and followed it rigorously. Three months later he wrote that the allergic spots and pains in his hands, arms, and feet had disappeared; he felt fully recovered.

The second case history is that of István, who was twenty-one at the time he consulted me. He had asthma when he was six months old. That problem was resolved, but since the age of two he had had symptoms of neurodermitis on the neck, on the legs, and occasionally also on the mouth. The tests indicated allergy to lactose. Since István was a professional dancer and traveled a great deal, he couldn't follow a rigorous diet. Thus I looked for the cause of his allergy as the first step of my therapy. Testing with the dowsing rod, I found a possible cause when he was twenty-one days old. I sent a rebalancing information and then asked his parents to shed light on this finding. It turned out that his mother's milk went dry on that day and she could no longer breastfeed him. He was given synthetic baby food and was nourished on that, subsequently in combination with ordinary milk. I suggested homeopathic remedies, and István's allergies disappeared in the span of three to four months.

SPACE- AND TIME-TRANSCENDING HEALING: HOW IT WORKS

In my practice of remote healing, the steps remain generally the same no matter how diverse the patients or their problems. The first step is to pick up information regarding the patient's condition. The information I receive through dowsing indicates the patient's overall condition of health and also the nature of his or her complaint. This diagnosis can be as detailed as any obtained by examination at a doctor's office. I then seek to discover whether the patient suffers from a temporary malady or a chronic illness. I work to identify the cause of the problem, and to what extent it may be due to environmental influences, such as electromagnetic or geomagnetic radiation or pollution.

In the next phase I analyze and structure the information regarding the patient. I arrive at a diagnosis, and suggest a therapy. Depending on the nature of the problem, I may prescribe allopathic remedies, a specific diet, phytotherapy, homeopathic remedies, or healing by information. I convey the particulars of the treatment to the patient by phone or even via email. If the therapy doesn't call for conscious cooperation by the patient I can effect healing purely by sending information without him being aware of what I am sending, or even that I am sending it. Consciousness on the part of the patient is not a factor, as long as he asks for and accepts the treatment. Otherwise the reception of the healing information could be blocked.

Healing by information can be carried out over any distance. The information affects the patient's condition and the effect can be verified through subsequent tests. These can be carried out by the dowsing method or by conventional means.

TWO CONTROLLED REMOTE HEALING EXPERIMENTS

The first of the two controlled experiments I report here took place at a seminar of the Hagia Chora Association in the town of Hohenwart in Germany on June 3, 2001. It was conducted by Günter Haffelder, director of the Institute for Communication and Brain Research of Stuttgart. The experiment was witnessed by about a hundred and twenty seminar participants and was monitored both by Dr. Haffelder and a volunteer physician, Dr. Heidrich Treugut. It was subsequently reported in the journal *Hagia Chora* (no. 9, August 2001).

At the beginning of the experiment I asked the subject—a forty-eight-year-old volunteer, one of the seminar participants—to give me a verbal report on his physical condition. He reported: "Four years ago I turned to a doctor with a serious problem in my hands. I could hardly move my wrist, and my fingers were swollen and in pain. The doctor diagnosed multiple arthritis. I decided to try a therapy consisting of a vegetarian diet and to study spiritual development. Now the joints in my fingers and my right wrist can move, but my left wrist is still stiff."

We then separated. The subject was seated in the main seminar hall in the presence of Dr. Haffelder and the participants, and I moved to a distant room with Dr. Treugut. Both the subject and I were wired with electrodes on our heads. Large screens in the hall displayed the electrical activity of our brains for the participants.

The experiment proceeded as follows. First I tuned my brain and nervous system for receptivity to remote information, and when I was satisfied of my own sensitivity, I began to examine the subject using the combination of the Körbler method and the method of healing by information I have developed. I examined his principal organ systems, and then his meridians. The colon indicated a mild irregularity, and I looked for and sent the appropriate healing information. Then I found an indication of inflammation on the left wrist, and for this, too, I sought the pertinent information. Then the liver meridian called for correction. Finally I examined the subject's pancreas and corrected a mild malfunction there. When I found that no further meridians or organ systems needed correction, I reinforced the balanced energy state of the subject by applying Tibetan healing symbols. Finally I tested for the exact duration of the healing information and found that the treatment for the colon and the left wrist was to last 10 days, and that for the pancreas 6.5 days. These effects unfolded without any further input on my part.

Reporting on the experiment, Haffelder wrote,

> In this experiment, which lasted about twenty minutes, an exceptionally high delta activity appeared in the brain of the healer, indicating the transmission of nonverbal communication from the healer to the patient. The healer perceives the malfunction in the patient, balances the pattern, and sends it back in a transformed

form. In general the process of examination by the healer manifests strong delta activity in the form of a significantly higher deviation of the regular rhythm in her left hemisphere in the range of 3–5 seconds at periods of 3–4 seconds. Synchronized with the rhythm of this brain activity, delta and alpha activity occurs also in the patient. The delta activity shows that the information was received, and alpha that it was effectively integrated.

The delta activity that appeared in this experiment is typical for adults in deep sleep, while alpha activity occurs typically in a restful state, with closed eyes. (In the normal waking state external stimuli generate beta waves in a higher frequency range.) It's significant that I had my eyes open in this experiment and yet my brain displayed activity typical of deep sleep. Equally remarkably, the subject displayed the same EEG pattern, sitting relaxed but not asleep. He proved capable of receiving the information I sent from a remote location, even though there was no sensory contact between us.

In my decades-long practice I have used this kind of procedure when treating cases of acute inflammation, pains from injuries, and a variety of more serious maladies.

Essentially the same experiment was repeated a few months later in Stuttgart at Haffelder's Institute for Communication and Brain Research. The experiment on October 29 of the same year was monitored and subsequently documented by the spectrum-analytic method of EEG recording, the same as the earlier Hohenwart experiment.

Before starting this experiment, the test subject, Katerina, who was forty-five years old at the time, described her complaint. She had suffered from allergic bronchitis for the previous ten years, a condition that was especially excruciating in the morning hours. She had consulted a number of physicians and received a great variety of medications, ranging from steroids and antibiotics to homeopathic remedies. I realized that there was not much point in concentrating on the symptoms themselves: I had better look for the causes. Then she and I were wired with electrodes, and Katerina went to another room in the laboratory, while

I began the procedure for diagnosis in a room shared with the experimenters. The electrical activity of our brains was displayed on monitors and was also recorded.

Following a preliminary examination, I proceeded to move back in Katerina's life until I found a particular trauma that could account for her condition. This event occurred in the immediate postnatal period. I administered healing information for the tenth minute after birth, and this process lasted 2 minutes and 41 seconds. Exact timing proves essential for the effectiveness of the healing. The healing information must be precise—a difference even of a few seconds can render healing ineffective, or produce undesirable effects. And it must be focused on the precise moment when the trauma occurred in the life of the patient. In the case in question the information I sent treated the trauma that occurred seventeen minutes after Katerina's birth, and the healing information lasted 45 seconds.

During the time that I carried out the examination and sent the healing messages, my brain exhibited EEG waves in the low delta region. Katerina's brain replicated my wave pattern with a delay of about two seconds. The effect was clearly displayed: as I was sending healing information she exhibited an aggravated form of her symptoms, coughing violently. When we concluded the experiment, the coughing subsided and Katerina calmed down.

On May 23, 2002, Katerina wrote: ". . . concerning my cough attacks, there were [after the experiment] some quieter periods and some periods in which I had violent coughing up to eight hours a day. Now it's quieter than ever before. The coughing didn't entirely disappear, but it's within tolerable limits. I haven't had such a quiet period as now for the last ten years."

My experience shows that it's possible to receive information on the condition of a patient over any distance, and from his present as well as from his past. This for me is significant evidence that a nonlocal information field mediates the exchange of information between healer and patient. The Akashic field is not an abstract theoretical concept but a working reality.

NINE

The Uses of Akashic Information in Business

William Gladstone

William Gladstone is anthropologist, author, filmmaker, literary agent, and founder and owner of Waterside Productions Inc. He received his undergraduate degree from Yale College and his graduate degree in cultural anthropology from Harvard University. Gladstone is the author of three published nonfiction titles and a novel, *The Twelve*. He is a director of the Club of Budapest, served as North American spokesperson for UNESCO's Barcelona Forum 2004, and is one of the founders of GlobalShift University.

I am neither a scientist nor a spiritual teacher. I have, however, been aware of receiving time- and space-transcending "nonlocal" information almost from birth. These Akashic experiences have shaped my life, and especially how I approach business relationships, business meetings, and business planning. Here I provide an account of how information that's beyond the range of our senses has contributed to success in my career as a filmmaker and literary agent.

Already as a child, I was aware that my mother had remarkable psychic abilities. We didn't have a name for her strange "sixth sense," and we didn't draw attention to it. In most instances it was almost taken for granted that she could sometimes predict what might happen to us or to our friends.

Among other things, she had an uncanny ability to get to know people who would later come into my life quite unexpectedly. She was in Europe with my father and met a young man and his girlfriend at a restaurant in Paris. She and my father enjoyed being with young people, and they befriended this couple. They took my parents to the latest disco clubs in Paris—so they could meet the "in" crowd—and in exchange my parents picked up the tab for the clubs and meals for the three days they shared company. Six months later I was sailing back from my junior year (of high school) abroad on the SS *Aurelia,* a low-cost ship that catered to students. On that trip I had become friends with a college student, and, purely by chance, he was standing next to me at the time the ship docked in New York. My parents were waiting for me on the dock and waving to me. My new friend was shocked when he saw them. He had been the young man they had befriended in Paris.

The most dramatic instance of my mother's sixth sense occurred on a different trip to Europe, when I was much younger. She woke up in Paris in the middle of a dream with a premonition that the boiler in the basement of our house in Scarborough, New York, was about to explode. She was so certain of her premonition that she called the house sitter who was taking care of me and my three siblings, all of us under twelve years old, and insisted that the house sitter call the fire department and have us all the leave the house. With the time zone difference, it was only about 9 p.m. in New York. When the fire department showed up and checked the boiler they determined that, had it not been fixed, in another hour it would have blown up.

I must have inherited some of my mother's ability to access such time- and space-transcending "Akashic information" for in my business life I have often had hunches or premonitions that not only turned out to be true but also extremely useful in the context of my professional dealings. I cite a few of these instances here by way of example.

AKASHIC INFORMATION IN MY SEARCH FOR ANCIENT MYSTERIES

As a young man of twenty-two, through some unexpected coincidences, I was offered the opportunity to serve as the researcher and production coordinator for the television special *In Search of Ancient Mysteries,* narrated by Rod Sterling, the creator of the *Twilight Zone* television series. This television special was a sequel to the Erich von Daniken television special, *In Search of Ancient Astronauts.* Both television specials aired in the early 1970s and were the most successful documentaries of that decade. Each was rebroadcast twice a year throughout the 1970s and led to the successful television series that kept the name, *In Search of Ancient Mysteries.* The series was narrated by Leonard Nimoy, who played the part of Dr. Spock in the original *Star Trek* television series.

It was while working on this documentary television special that I first became aware of the power of Akashic information to assist me. I had done research on the ancient mysteries the film treatment called for, searching specifically for as many "ancient astronaut" images and references as possible. This included the Nazca lines and other legitimate mysteries that could be tied to potential theories of extraterrestrial contact, but others turned out to be completely bogus references. Instead of panicking when I would discover that a rock formation that was supposed to resemble ancient spacecraft only had this aspect in the month of March under specific weather conditions, I would simply open my mind to the possibility that, having flown a film crew with equipment halfway around the world, we might find some genuine "ancient mysteries" to film. Inevitably we did. I had a feeling that somehow the universe itself sent me the right cues, so the film could demonstrate that there were ancient technologies of inexplicable achievement—such as brain surgery being performed in sixteenth-century Peru and the creation of edifices such as Sacsayhuaman outside Cuzco that defy present-day building techniques.

FURTHER AKASHIC INTUITIONS

In my subsequent business career I continued to experience hunches and intuitions that proved extraordinarily useful. As a literary agent and head of Waterside Productions, in the six months between March and September 1983 I sold more than fifty computer "how to" titles and was quickly established as the most dependable source in this technical field. This created something that I now look back on as a veritable vortex of energy and intuition. At least twenty times during that period I received a phone call from a publisher or author seeking a specific technical skill, or a book project, and within five to ten minutes the ideal person to fulfill that request would call me or sometimes unexpectedly show up in my office! These intuitions were reminiscent of what I had experienced in filming *In Search of Ancient Mysteries*. They were extraordinarily productive. Although at the time I had only one part-time assistant, in a single twelve-month period just two years after entering in the field of computer book agenting, I sold 299 titles with great assistance from these remarkable coincidences.

In my early years as literary agent authors often called me to ask for advances on their contracts, since publishers could take weeks if not months to pay and the authors were sorely in need of cash. In such instances I would ask my assistant if we had received any money we might use to prepay the author. More often than not she would say no. I would then say just go ahead and write a check, because some money will arrive right away—and even if it didn't, we could just hold the check. Not once did she have to hold the check, as my intuition proved right and the money would arrive, sometimes that very day, even in instances in which it was not actually expected.

My business activities continued to benefit from such Akashic information. A skeptic might say that I have been extraordinarily lucky in my career as literary agent and filmmaker, but I would ask, "Yes, that may be true, but what lies beneath such luck? Might it not indicate processes at work that are best understood as Akashic experiences, in this case taking the form of spontaneous hunches and intuitions that work out miraculously?"

I have become convinced that we must remain open to all the information that's reaching our brain and mind, regardless of whether it's coming to us through eyes and ears or through more subtle yet just as effective channels. Surprising as it may appear on first sight, even business-related information may come to us through the unfathomable storehouse the ancients knew as the Akashic Record and what Laszlo connects with cutting-edge science as the Akashic field.

TEN

Visiting the Omniverse Center

A MIND-TRANSFORMING AKASHIC EXPERIENCE

Oliver Markley

Oliver Markley holds an M.S. in engineering from Stanford University and a Ph.D. in experimental social psychology from Northwestern University. He is professor emeritus and former chair of the graduate program in Studies of the Future at the University of Houston-Clear Lake (UHCL); prior to that he was a principal investigator and management consultant with the Management and Social Systems Group at the Stanford Research Institute (now SRI International). A fellow of the World Futures Studies Federation, Markley is the author/coauthor of four books and more than fifty other publications and has been a consultant for various organizations in diverse sectors of society.

Although my first Akashic experience dates back to when I was three years old, these experiences began in earnest when I was a student and, later, a professional colleague of the late Willis W. Harman. One of the most phenomenal experiences that my professional path opened up for me was to encounter an extraordinary Akashic repository of knowledge and skill, which called itself "The Omniverse Center for Cultural Development."[1]

THE METHODOLOGICAL CONTEXT OF THE "OMNIVERSE" EXPERIENCE

In 1969, as a fresh postdoctoral professional, I was assigned the task of leading the methodology development at the new futures research think tank that Harman was founding at the Stanford Research Institute (now SRI International). As if it wasn't hard enough to figure out how to do holistic research on alternative futures of society (possible, probable, and preferable), our first major results indicated that of some fifty of the most highly plausible alternative future histories for society, only a small handful were by any stretch of the imagination desirable, and most of them involved deep-seated transformation regarding underlying attitudes, images, and policies in response to emerging problems involving overpopulation, resource depletion, pollution, dangerous weapons buildups, and so on—problems that Harman collectively dubbed, "The World Macroproblem."[2]

With my methodological responsibilities in mind, I in turn reasoned that research methods based on rational/analytic modes of thinking are, in principle, not suitable for creative exploration of transformational alternative futures because such thinking modes are, at root, essentially mechanistic extrapolations of what has gone on before.

My knowledge of cognitive psychology pointed toward *intuition* as the appropriate mental mode to use, so I gave myself the task of searching out as many different ways to access one's intuitive faculty as are available.

By the time I experienced the episode described below, I had already tentatively selected *guided cognitive imagery* as the most appropriate technology for helping people tap their intuitive sources of knowledge. But I was aware that I had not yet checked out the method of "trance mediumship" (more frequently called "channeling" these days), but I felt some reluctance to do this because of the rather unacceptable image that mediumship often has for people with a scientific orientation. However, since open-mindedness to alternatives is a central characteristic of good futures research, I felt I owed it to myself to at least see what this modality might open up.

As things turned out, the following experience—and other things it led to—answered all the questions I felt I needed to pursue at that time regarding mediumship or channeling.

THE CONTENT AND PROCESS OF THE "OMNIVERSE" EXPERIENCE

One afternoon in the spring of 1976—just about a week before I intended to find a trance medium to visit—as I was riding my bicycle from my SRI office in Menlo Park to my residence in Palo Alto (a distance of some 2.8 miles, through mostly residential streets), just as I left the SRI property line, I heard a voice saying,

"Hello, I'm Henri. I'd like to show you something. Would you be willing to come with me?"

My first reaction was to look around to see the source of the voice. Finding none, I realized that it came from inside my own being, as in "hearing voices."

My second reaction was one of interest in the fact that because I am not conversant in French, I had no way of knowing how the name Henri was pronounced and wouldn't have understood its sound if I hadn't simultaneously been given to understand its correct spelling by means of some type of nonverbal intuition that was immediately obvious.

This being the case, I at once concluded that this must be some type of experience involving "higher consciousness," and that although there was no one I could "see" who had spoken to me, it probably wouldn't be a bad idea to play along with the situation and see what happened. (After all, because of contextual factors described above, I had been planning to soon begin an investigation of trance mediumship, and this experience seemed to be somehow related to this intention.)

So I thought, "Sure, I'd be glad to go with you."

For the next several moments we seemed to be under way to some unknown destination that I was silently given to understand was "beyond space and time." Meanwhile, I reassured myself that I was functioning normally, with clear awareness of what was going on outside. I could, in other words, continue riding my bicycle safely, even though some new type of mental experience was going on inside.

The next thing I was aware of was that we had somehow stopped our journey to wherever and whenever our destination was. Where were we? I didn't know. Nor did Henri give me any clue. In the unmoving silence, I somehow intuited that the next step was up to me.

This being a totally novel experience, I asked myself, "What knowledge or prior experience do I have that would apply here?"

Two phrases from the occult literature immediately popped into my mind: "Guardian of the Threshold" and "Initiation." With this to guide me, I composed the following thought, which I then silently "transmitted" to who or whatever was in control of the experience I was in the midst of: "I don't know where I am or why I am here, but I didn't ask to come—I was invited. However I can say this: Although I don't know what this is all about, I will either take responsibility for using whatever I find here, or I will not use it at all."

At once, it was as though an invisible shield was removed, and I found myself mentally gazing on a city of incredible beauty floating in midspace immediately in front of me. At the same time, I was given to understand that the city I was seeing was but a three-dimensional spatial metaphor of a multidimensional reality, a personal "projection" that was necessary if I was to perceive what was here in ways that would be meaningful to an earthling like me.

As we entered the city, I noticed a building on the right side of the street we entered. On the side of its foyer was an opening that looked like the hatcheck window of the Hollywood Palladium I had known as a youth in Southern California. In this window was a swarm of lights, which looked very much like tiny clear white Christmas tree lights. But each light was obviously a sentient being, and the whole swarm was also sentient in its own right.[3] The swarm thought to me, "*Welcome.*"

I thought back to the swarm, "Thank you. Where am I? What is this place? Does it have a name?"

They thought back, "*You (or, your species) would call it the Omniverse Center for Cultural Development. It is an intellectual oasis for evolutionary operatives such as yourself.*" (To clarify, I should add that the word *omniverse* in this context clearly meant "all creational epochs, past and future.") Although a paradoxical concept, it made complete sense as they thought it to me. And the word *oasis,* in this context, had similarly multiple meanings that now, many years later when knowledge of the Akashic field has become available to me, make much more sense.)

I was at once overjoyed that something like this was possible and that an experience of it was being offered to me, without my even having had the wit to ask for such a thing.

I thought back to them, "What is your function?"

They answered, "*We provide a guidance and translation service so that visitors such as yourself can get what you need here.*"

I thought back, "What benefit do you get from this arrangement?" It was at this precise moment that I became convinced that what I was experiencing was a valid experience, for as I asked this question, a ripple of mirth swept like a wave through the swarm as they replied, "*We get all [the information/knowledge/intelligence/wisdom] that passes through us for our own use.*"

By this time it was evident to me that I was addressing an entire population of telepathically sentient beings whose niche in the multidimensional ecology of consciousness that the "Omniverse Center" linked into was to help visitors get whatever they need.

My next response was predictable: "What is there to see here?"

Rather than see a menu or map, I was at once treated to a rapid series of lucid impressions that closely resembled what computer operators call a "core dump." Although much too fast to permit me to dwell on any one impression, I nevertheless had a good sense of the whole, most of which was well beyond anything I would have thought possible or even conceivable.

I next asked, "What should I look at first?" But they refused to advise me, indicating that I had to make up my own mind what I wanted to see or experience.

At this point, I drew back inside myself and reflected on the fact that historically on Earth, when a relatively more "advanced" culture came into contact with a more "primitive" one, things usually didn't go too well for the primitives. So I ventured to ask, "What is the ratio of war and peace in the Omniverse?"

I was guided down a hall to the left, into a door on the right, and stopped in front of a device that had a television-like view screen. Soon the screen warmed up and displayed a cluster of lights that were connected by a network of lighted threads. Some of the lights were white; the others were red (the significance of the colors was immediately apparent). The ratio was about 2/3 white to 1/3 red. I asked, "But what is the trend?"

After a few moments, a deep psychic voice (which *felt* to me like the voice of "God, the Father") came down from above, saying something like, "*We judge that you* [that is, your intellect, degree of mental development, and so on] *are not sufficiently developed to understand a meaningful answer to your question. Nevertheless, if you insist, we will find a way to answer it as best we can.*"

I deferred. After looking at a few other things, I realized that I was by now within several blocks from my residence, and since I didn't want to greet my wife and children with this sort of experience going on, I began my exit from the Omniverse Center. Henri was waiting for me as I left. He asked, "*Do you have any questions?*"

I replied that I had only one: "Can I come back?"

He said, "*You can come back any time you need to.*"

That evening, after my wife and family had retired, I sat in the reclining chair I kept in my home office, let myself become relaxed into a nonlocal level of consciousness, and then checked to see if it was possible to return to the Omniverse Center. My intention was to ask about a technical problem I thought might be a suitable way to explore the use of this new tool.

Upon opening myself to this possibility, it was immediately present in my consciousness once again. I inquired about the question I had brought with me, got a satisfactory answer, and let my consciousness of the Center go. I then noted the answer in my journal and went to bed.

The next day at work I mentally returned to the Omniverse Center and got an answer to yet another question, but when I tried to return later that week, it was not available to me. Instead, I found myself recalling what Henri had said when I had left the first time . . . that I could  return whenever I *needed* to. Concluding from this that this exciting resource was not to be used frivolously, I mentally filed the idea that the Omniverse Center would probably remain as a resource to be used whenever *truly* needed, but what would qualify as valid "need" in this regard was probably knowable only by intuition and not by superficial rationality or opinion on my part.

CONSULTATIONS WITH THE OMNIVERSE CENTER

During this general time period, I regularly engaged a small group of my SRI futures research colleagues to help me pilot test the variety of guided visualization procedures I was developing to foster the use of intuition as a means to explore the future. Our method of operation was simple. Each Friday, in the early afternoon, we would gather at the Atherton residence of Arnold Mitchell, the founder of the SRI VALS (Values and Life Styles) Program. While having light refreshments, we would talk about our ongoing research projects with an eye toward the selection of interesting research questions that might be illuminated by the visionary/intuitive procedures I was developing. After selecting several interesting targets for investigation, I would intuitively fashion specific methods to use in our afternoon's work and we would begin. (By way of illustration, in one session we decided to explore possible smog levels in the future. After first using a type of relaxation and mental focusing useful for putting us in the right state of "nonlocal" consciousness for this type of visionary/intuitive exploration, we each imagined crawling into the same ten-foot-diameter eyeball—what nowadays would be called a "virtual space-time ship." Once inside, we would collectively fly to various space-time locations, such as over the Los Angeles basin at an altitude of 2,000 feet in the year 2020. We would then compare notes on what we saw. The results would then be used in the "regular" futures research we did back at the office the following week.)

On one particular Friday afternoon, two researchers in our group reported difficulty in writing a scenario involving "the Man on a White Horse" (a social science phrase that refers to the phenomenon of a charismatic leader who revolutionizes a society, often by finding a scapegoat to blame for current societal problems). They simply couldn't get the scenario to "work," and therefore asked if there was some visionary/intuitive method that might resolve their problem.

As is my custom when getting such requests, I "went inside" to see what my intuition might suggest by way of an exercise. (From almost the beginning in this type of work, I have found that guided visualization works best when the guide is guided by his intuition rather than following some prescribed script or program of instructions.)

My intuition immediately suggested "the Omniverse Center" as an avenue in which their questions could be appropriately answered. This greatly surprised me, for I had never considered the possibility of taking others to a "place" in consciousness lying outside of normal time and space, and I wasn't even sure how I could do it.

My reluctance quickly gave way to interest when my intuition also suggested playing the "Night Music" from Bartok's *Music for Strings, Celeste, and Percussion* as background for the exploration, for this particular section of music has often been instrumental in helping me facilitate entry into interesting states of consciousness. So I cued it up and told the participants that if they were willing, I would lead them on a new type of visionary exploration—one that might even seem like being outside of time and space but not to worry, I had been there before and would guide them into the experience.

I then used a conventional relaxation induction and suggested that we would get to our target by "flying in formation" like air force pilots do, with me in the lead; that I would take them to an interesting place; and that once there, they should do anything they felt appropriate in order to gain access and get meaningful answers to our questions. (Three people volunteered to inquire about the Man on a White Horse/

charisma issue; the others simply went along for the ride, with instructions to bring back anything they found interesting.)

Our attempt to "fly in formation" worked flawlessly, and once I was there, I "took muster," checking to make sure that the others were there as well. I then said that I would give them ten minutes to explore whatever they found there while I "stood guard," so to speak, at the entrance and that I would then bring them back to "the room" (our phrase for the normal state of consciousness from which we departed on our visionary expeditions).

The results were striking. The "fellow travelers" had experiences that, while interesting, were nothing to write home about. The three who went to inquire about charisma, on the other hand, came back so deeply moved that we had to take a fifteen-minute break in order for them to integrate their experience and stabilize their emotions. It turns out that each of the three came back from the Omniverse Center with what can only be called a "religious conversion experience" in which they experienced the reality of charismatic insight in their own lives and came to see how charisma is an essential ingredient in all valid leadership. They saw that in their research they needed to radically reconceptualize the scenario into a story in which charisma is used constructively, rather than destructively, in society.[4]

Without question, the Omniverse Center experience has unalterably transformed my *Weltanschauung*—my vision of reality—and it did so for those of my SRI colleagues who accompanied me there as well. Moreover, the information obtained there proved very useful in practical real-world terms.

ELEVEN

Singing with the Field

THE DAISY HEART OF SPIRAL CONSCIOUSNESS

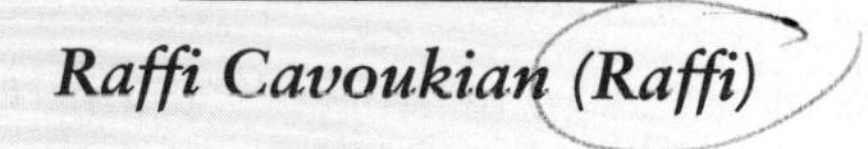

Raffi Cavoukian (Raffi)

Raffi Cavoukian, C.M., singer, recording artist, author, and ecology advocate best known as Raffi, is founder of Child Honoring and president of Troubadour Music. He is a recipient of the United Nations Earth Achievement Award and was named to the Global 500 Roll. Raffi holds two honorary degrees, is a member of the Order of Canada, the Darwin Project Council, the Spiritual Alliance to Stop Intimate Violence, and the Club of Budapest.

Me? Write a report on the Akashic experience? As soon as I pushed aside the doubt, examples of such experiences—extraordinary, unexplainable apprehensions outside normal perception—came to mind effortlessly, and I accepted Ervin Laszlo's invitation.

I had first heard of an Akashic Record in the spiritual books I was reading while taking yoga classes in the mid-1970s, and I found the concept fascinating. If there could be a collective unconscious (as Jung had said), why not a universal unconscious or *superconscious*, a huge reservoir of all that has been and will be? The mind boggles, and yet we can appreciate that a vast storehouse of manifested and unmanifested phenomena might somehow exist in the timeless, spaceless realm we call eternity, or God, or Source.

There's comfort in knowing that we can "tap" this Source for guidance, for knowledge and inspiration—by conscious invitation, prayer, or meditation. Where do ideas come from? Basic functioning thoughts might come from self-interest or self-fulfilling drives of our physiology. But creative ideas perhaps reach us from elsewhere, from the universal archive of infinite possibility.

The very writing of these words has just now turned into an experience with Akashic overtones, as with eyes closed I type what multisensory impulses move my fingers to do. A pause to breathe, reflect, breathe. And I'm typing once again.

Where do musical ideas come from? I've often felt pregnant with a song idea that came knocking on my consciousness and, once "let in," continued to grow. Often this leads to the idea gaining fullness and birthing as a song. At times not. Some song ideas don't make it through the crowded contours of my daily life.

Since I began meditating in 1979 (steadily, if not as regularly as intended), I have been a successful songwriter devoted to the greatest good, with my creative ego in service of the whole. Interestingly, I can't recall any "writer's block," ever. Through a sustained channel of appreciative inquiry into the human and Divine, I'm flooded with ideas. (All the more reason to sit quiet, to bathe in the Akashic realm, in universal mind.) If anything, the number of song fragments and essay themes streaming in require more time and consideration than there are hours in a given day or week. This creative abundance is a great joy and, only sometimes, a burden.

GRACE WITH A CAPITAL G

The Akashic experience comes in different hues and sizes. Messages and visions of varying length and clarity seem to go with the territory. In my twenties in Toronto, while sipping a beer at an outdoor patio and stroking my black beard, I looked up and a horizontal string of lightbulbs (from an adjacent property) arrested my attention as if to greet me, one by one, turning on a light in me—a cue to leave all cares behind. In the glow of that tranquil moment, which stretched on for a while, I faintly recall paying the bill and walking to my flat in a state of quiet levity and humbled respect. During the short walk street noise abated and lights almost hummed their presence. A profound ease held me aloft as, one foot in front of the other, I made my way. I was in a bubble not of my own making, but of my own accord, acknowledging the specialness of this feeling even while I wondered how long it might last. Unhurried, I breathed this nuanced gift of time, inhaling and exhaling the non-temporal state. This was the first of such experiences, and it came well before my spiritual journey had consciously begun.

During challenging times in my life, vivid dreams have held messages and confirmations. One very clear dream brought a palpable form of Grace. In the dream scene she entered, I was lying on the floor with friends watching a revolving LP album whose center label suddenly became three dimensional and floated upward like a beautifully colored orb slowly spinning in the air. In that delicious moment—with all of us very relaxed and just smiling at the magic—that unmistakably said ALL was profoundly well and, YES, without a doubt there *is* a God and no cause for fear or worry, in walked the personification of Grace, a beatific brunette, smiling as she reached over and shook my hand. My God, what a sweet messenger.

And some years ago in my fourth-floor Vancouver apartment, I was visited by an invisible Grace, with a capital G. I didn't know what else to call her, as "it" just didn't fit. Out of nowhere and everywhere she came, a vibration, a feeling, a knowing that "all is profoundly well, and will be well." An emotional and physical assurance. A soul massage of every care as my shoulders dropped and face relaxed, lips opened with an ever so faint smile at this slow-motion rapid response to Grace. If I sensed what (or who) this was, it's because at several points in my life I'd had such curtain raisers, peak experiences that fit in no known file or category. By the time of this episode, I was more than ready. It didn't faze me at all that some of the framed paintings on my apartment wall signaled they wanted rearranging. I smiled and made two paintings change places and was pleased with the result. Grace moved me to bow to these inheritances from my departed parents who had exited this world dramatically—from different causes, in the same hospital, within twelve hours of each other. Their memories danced again in my living room.

To my ongoing delight, Grace lingered for about two hours. I gathered myself and drove to work, every step and movement conscious and steady even in my exalted state. I entered the workplace, called my assistant into my office, and explained what was happening. With understanding and sensitivity she nodded and closed the door to leave me with my thoughts. As the experience faded, even during lunch there were traces of this magnificent humbling feeling superior to most I've ever known. Quite unexpectedly, there was a shorter visit some months later, but she hasn't been around since. Still, Grace has appeared to me in different guises.

A VISION ONE SUNDAY

In 1997, a vision came to me, startling, yet not altogether a surprise. After years of attuning my mind and heart to the world of the very young—as a popular family entertainer with songs for children, and considering the factors that shape human development on our planet—it came to me.

At 6 a.m. on that Sunday morning, from a sound sleep I sat upright in bed with eyes open and "saw and heard" the words *child honoring*, crystal clear. I understood that Child Honoring, a unique philosophy at a critical time in history, for all humanity, is based on the idea that how we regard and treat our young is the best way to create humane and (therefore) sustainable cultures. And I felt in that moment that my whole life thus far had led me to this revelation, and that it would be the work of the rest of my life. ("What a relief, to know one's destiny," I thought.) And in that moment I knew intuitively what would then take me years "to know that I knew," to conceptually develop and unfold the enfolded treasure of that instant. In wide-eyed reverie, I realized that I was being given a philosophy that linked the personal, the cultural, and planetary domains as never before, with the child as inspiration. I could not recall a social revolution in history inspired by the growing child, the universal being of all cultures. I trembled with excitement, and some fear. And still, I heard in my mind that though I might be ridiculed for what some might dismiss or deride as simplistic, I would not waver in advancing this philosophy.

I knew right then the importance of connecting the dots between how we parent, the cultural values we transmit (knowingly and unconsciously), and the planetary conditions that make up physiology. They all affect behavior and development. I could imagine and intuit the enormous changes that would come if society understood the critical importance of early years in the shaping of *developmental intelligence*—the key to building what some call "social capital." I knew that with this knowledge writ large, all our institutional values would have to change. We would move heaven and earth to protect and support our most valuable players: the infants whose dazzling brains and innate brilliance are in our care.

Now and again I marvel at the fullness of that one moment's knowing, given to a Cairo-born boy of Armenian parents who came to Canada at age ten (in 1958) and grew up to sing and dance with the children and families of my adopted country and its neighbor to the south. And here's what else I knew in that moment: the very phrase "child honoring" brings forth questions in the hearts of those who hear it. It starts with, "Were you honored as a child for who you felt you were?"

For some years previous to this, I had used the hyphenated "child-honoring" in essays as an adjective. And now the words came rearranged to name an integrated philosophy. I could not contain my excitement, and God knows that I have been overbearingly single-purposed ever since, much to the disappointment of fans who wanted to hear me sing the old favorites, and to the chagrin of friends who sometimes just wanted to talk about something else! While I'm far more able to make small talk or engage a number of serious themes, the child as inspiration continues to dazzle my dot-connecting mind if for no other reason than because the child doesn't live in isolation but in the circle of all our relations; she touches everything that's meaningful to us. Oh, and one other reason: because we can view the child as a quantum being—a state of the possible!

THE CHILD HONORING CRYSTAL

In some measure, I believe that the child as universal human, in whom an essential humanity is most visible, brought me the Child Honoring "crystal," as the vision increasingly came to reveal itself. As a crystal, every face and facet of Child Honoring reflects and refracts the others. How you look—how you hold the crystal in your innermost being—affects what you see, and where and how you look shows varying vistas depending on the angle (or bias) of light and mind and heart. Luckily for me, a number of close friends became intrigued with the crystal image, and their embrace of it illuminated the many-chambered lens of this multifaceted philosophy. The child "seen, heard, and respected" became a lens and metaphor for seeing and knowing all that's precious in being human, and in our planet, our habitat, our home.

Robustly we engaged Child Honoring—critics and admirers, we jostled and squeezed the theme and its principles for every ounce of meaning. We argued, laughed, and cried about what all this might mean to self and to society. In time, the question "Were you honored as a child for who you felt you were?" moved many to tears as they confronted the far-from-honoring trauma of early years. It stirred me to imagine a world where such pain is not the norm.

The Akashic revelations of that singular day have served well over the past ten years. In settings as diverse as Harvard, Ottawa's Parliament Hill, and the New York Academy of Medicine (and at campuses across Canada and the United States), the philosophy has been enthusiastically received. The words Child Honoring cause all manner of emotional reactions in people, depending on their personal growth and emotional awareness. A great number have felt deeply touched (and troubled) by the look through the lens. Some parents felt new waves of guilt over not having parented well enough, despite repeated assurances that this work is not about guilt. Joyful and affirmed responses came from those who felt they now had a meta-framework for a paradigm shift that made more and more sense to them the more they engaged it. And yet in many, I saw a distancing from children and childhood, as if early experience is some foreign field and not the one we all played in.

At times I've struggled to hold the breadth and scope of the philosophy entrusted in me to bring forth a vision not mine but hopefully destined to be a shared dream, an idea whose time has come. Why me? I have asked. Why was I given this vision? I'm neither a parent nor a certified educator; I'm a singer with an inquiring mind and an open heart. Is that enough? Was it something about my appreciation of children and their pure love that "got the call"? Was it my standing as a troubadour, a voice free and unbound to any party in the status quo? Perhaps it was the joy and love of millions of children who sang my songs daily in the past three decades. Or the ecology advocacy that has been as much in my consciousness as anything in my adult years. However the Akashic impressions rang out, it's a humbling feeling to have a grand epiphany in whose crystal vibrations one senses a paradigm shift of Copernican scale. And perhaps in the nick of time for an endangered humanity on a planet in peril.

As a photographer's son, I had been dazzled by the first NASA photo of our blue-white Earth from space. What an unusual "group portrait"! Something unpredictable, and utterly compelling. Home, the whole of Mother Earth, now visible to all her children. And what conflicting factors combined to give us this archetypical image: explosive rocket science and advanced photography yielded not only the optimal distance needed to view the whole Earth image but also the means to record the feat. As with the Internet technology that emerged from military sources, an iconic image for evolving consciousness arose from the conquistador fossil-fuel age humanity must outgrow if we are to have a viable future.

For decades children and ecology had been the double helix that grew and danced within me as I sought to understand human development, evolution, and the numerical basis of the manifested sight-and-sound world. The world as vibration inspired me again and again. It stirred renewed appreciation for the magic of Creation. In my home on a small island off Canada's west coast, magic is abundant and tangible.

DAISY, DAISY

What possible photographic image could convey the breadth and depth of a philosophy whose name bore the word *child* and yet was about much more? What image could capture the Child Honoring lens, the crystal design of being and belonging? The answer came as a delightful surprise: the daisy. I photographed one in my front yard, and its simple beauty graces the front cover of the anthology that was published in 2006.[1] The daisy heart's spiral designs—what some call sacred geometry—seemed a perfect fit for the mystery and grace of being born into this field of wonders.

It was superb, as epiphanies go. I was given a novel lens containing a mandala of the universal human heart—the child heart as a very strong attractor to all we hold dear. Past and future came together in the luminous present, full of possibility. I can say that my life (past and future) flashed before my mind's eye and left impressions for future moments to activate. Smaller epiphanies from time to time bring succeeding levels of understanding.

How blessed I feel to have awakened that Sunday morning, with a new song to sing, with a new lens for seeing and a daisy for remembering. How does a daisy (and her big cousin, the sunflower) get her spirals? In Gyorgy Doczi's book, *The Power of Limits* (1981), we encounter spirals and the numerical ratios that bring them to life. This exquisite book reveals nature's proportional, reciprocal designs and the mathematical basis underlying their form. It was an Italian scholar, Fibonacci, who discovered the sequence of numbers that, expressed as ratios, described the curve of any spiral. (The Fibonacci ratio played large in Dan Brown's *The Da Vinci Code.*) Once aware of the presence of spirals in the world, you see them everywhere. And the little daisy with its spiral core can mesmerize you, as it did me.

One afternoon on a small island off Canada's west coast, a daisy surfaced to bring unexpected magic. Holding it was a five-year-old girl, Emily, who was present among a small crowd gathered to hear me present Child Honoring. Near the end of my word and song presentation, during the song "Turn This World Around" in which I was singing "Turn turn turn, turn this world around, for the children . . ." Emily did something totally on her own, without any prompting. She brought a daisy into the hall and went around giving its petals to those present. She gave the daisy's spiral heart to a woman who was pregnant.

When what I thought I was seeing in front of me was later confirmed by a mutual friend (as well as colleague), I was in awe. This little girl had been sitting with her father in the third row from the front for the first half of my presentation, and then with the start of my singing she had wandered about, perhaps stirred by the rhythms of the recordings I was singing with. Quite possibly she saw the Child Honoring anthology (with the daisy image cover) for sale at the back of the hall and wandered outside and returned with a fresh daisy in hand, one she wanted to give away. Pure coincidence, some might say. And yet for me, astonishing synchronicity. She acted with connection and knowing way beyond her tender years. Had she possibly tapped in to the vibrational field of the subject at hand?

No doubt the same Source field that woke me up that fateful Sunday moved a little girl to share her daisy with exquisite timing. What a dazzling game! The unfathomable mysteries of life on occasion reveal parts of the holographic universe in which each of us has a whole part to play. Ordinary perception bows to extraordinary insight with glimpses of unbounded multidimensional awareness in which we live and grow, emotionally and spiritually. My experience of God or the Akashic Source is humbling, awe-inspiring, and more so with each new flash or "aaah!" on the spiral rings of consciousness.

As I get older there is a feeling of both temporal time and the eternal at once, the constant witness living the witnessed life. The galactic scale of our vast and spiraling Milky Way and the billionfold plenum all around boggle my mind to quietude. Every now and again, a state of grace.

TWELVE

Connecting with Universal Mind in the Creative Process

Alex Grey

Alex Grey is an artist specializing in spiritual and psychedelic art (or visionary art) that is sometimes associated with the New Age movement. His oeuvre spans a variety of forms including performance art, process art, installation art, sculpture, visionary art, and painting. Grey is a member of the Integral Institute and is also on the board of advisors for the Center for Cognitive Liberty and Ethics and is the chair of Wisdom University's Sacred Art Department. He and his wife, Allyson Grey, are the cofounders of the Chapel of Sacred Mirrors, a nonprofit institution supporting visionary culture in New York City.

In 1976 my wife, Allyson, and I had a simultaneous vision, a mystical experience, which changed our lives and our art. We sacramentally ingested a large dose of LSD and lay in bed. I entered an altered state of awareness where I could no longer sense physical reality or my body in any conventional sense. I felt and saw my interconnectedness with all beings and things in a vast Universal Mind Lattice. Every being and thing in the universe was a fountain and drain of love energy, a cellular node or jewel in a network that linked omnidirectionally without end. All duality of self and other was overcome in this infinite dimension. I felt this was the state beyond birth and death, beyond time: our true nature, which seemed more real than any physical surrounding and more real even than my physical body. The matrix arose out of a field of pure emptiness.

After I opened my eyes to behold Allyson and our bedroom once again, her report of the experience exactly replicated mine. She described the same transpersonal dimension in her recalling and in her drawings of the awesomely vast and beautiful state of being. Experiencing the infinite net of spirit transformed our lives and the subject and focus of our art and mission.

Mentally consumed by this experience, I searched many references and found near-death research and mystical literature that related powerfully to the revelation. The Hwa-Yen Buddhist description of the Jewel Net of Indra was one such reference:

> In the abode of Indra, Lord of Space, there is a net that stretches infinitely in all direction. At every intersection of the net there is a jewel so highly polished and perfect that it reflects every other jewel in the net.

All of the work we have done since refers to the insights from that transcendental state and related spiritual experiences. Allyson began a lifetime of painting sacred geometric gridlike Yantras. She titled two seminal works *Jewel Net of Indra.* I started a series of paintings called the *Sacred Mirrors,* which featured views of the *Universal Mind Lattice,* a painting portraying a transpersonal latticework of living light bodies, the geometric infrastructure of interconnected beings.

The *Sacred Mirrors* are twenty-one paintings that function as one piece, exploring who and what we are, revealing layer after layer of material, biological, sociopolitical, subtle, and spiritual aspects of the self. Because they are paintings, the viewer stands in front of each one in turn and is invited to reflect on and identify with the systems portrayed, such as the *Nervous System* and the *Cardiovascular System.*

The various physiological and anatomical systems ground the viewer in the known physical world.

The inspiration to create the *Sacred Mirrors* series was catalyzed by Allyson, who suggested that I provide a physical-to-spiritual map to reawaken the viewer's connectedness to these systems. Psychedelic experiences inform certain paintings, like the *Universal Mind Lattice, Psychic Energy System,* and *Spiritual Energy System.* The overall vision of the *Sacred Mirrors* suggests a transpersonal point of view not tied specifically to one faith, but pointing to the wisdom that runs through all sacred traditions.

A VISIT TO THE AKASHIC RECORDS

Once, after taking LSD, I visited the Akashic Records. It looked like an x-y-z matrix of infinitely long hundred-meter-square hallways or corridors. The walls were made of a living iridescent stone and inscribed with many symbolic scripts, like hieroglyphs and pictograms, mathematic and physics diagrams with every known and untold alien language. Everything that had ever happened, was happening, or would happen was written there like a person's genetic code. Yet this was the historic or memetic code of planet Earth and its relation with other planetary, galactic, and interdimensional neighbors. Though the walls were inscribed, their living nature seemed to accumulate wisdom, to modify and improve, to cross-reference and update, like a Google search engine.

As vast and dense as this alternate plane of existence was, I was sure this was just the local branch of a library that was an endless and interconnected grid. I felt an overwhelming unknowability, an impossibility of reading the record, of ever really comprehending the profound mystery of the cosmos encrypted in this text.

As I got closer to the wall and looked at the symbols, I had an immediate understanding, as if the wall itself were a self-translating "overmind," a resource that any being could tap in to to gain insight or inspiration from the omniscient memory field of all that is. Creator as linguistic infinite intelligence, it transmits all wisdom texts and poetry. On occasion I have had the chance to hear, even if dimly, such transmissions. I have compiled a collection of these mystic rants in a book called *Art Psalms*.

The *Net of Being* painting recounts a visionary ayahuasca experience of infinite space composed of infinite Beings with infinite shared consciousness. A fiery web of eyes and galaxies form a grid of interconnected godheads throughout all space, establishing a new topology for the Self as continuum. The shared heart of each four-faced Quad God sheds an eerie underlight, as a ball of white light inside each head sources the luminosity for a level above and beyond sight, a mesh evocative of the mythic Jewel Net of Indra or Buddhafields from the *Avatamsaka Sutra*.

MYSTICAL CONSCIOUSNESS AND THE CREATIVE PROCESS

In an effort to illuminate the many stages of the creative process, I'd like to share something of the story behind my painting *Transfiguration*. I have always been mystified by the body-mind-spirit relationship and the difficulty of making these multiple dimensions of reality visible in a work of art, but not until my LSD experiences did I want to make mystical consciousness itself the subject of my art.

It took me about ten years of creating art and obsessively reflecting on this subject to reach the understanding that this was one of my primary artistic problems, an important part of my vision. I prepared a slide show and lectured on the subject of "Transfiguration," showing artistic representations of transcendental light or energy in relation to the body. At that point I didn't know that I'd be doing a painting by that name. There is an incubation stage in the creative process where the vast womb of the unconscious takes over, gestating the problem. The embryonic artwork grows effortlessly at its own pace. For the *Transfiguration* painting, this phase lasted about half a year. Then early one morning I woke from a dream. In the dream I had been painting a piece called

Transfiguration. The painting had a simple composition, two opposing spherical curves connected by a figure. Floating above the Earth sphere, a human figure, fleshy at the feet, became gradually more translucent. At about groin level it "popped" into a bright hallucinogenic crystal sphere.

The dream revealed a unique solution to my simmering aesthetic problem of how to portray the transfigured subject. But this illumination or inspiration phase, my "Aha!" moment provided by the dream, was extended or underscored later that week when I smoked DMT for the first time. As I inhaled the immediately active and extremely potent psychedelic, I got to experience the transfigured subject of my painting first hand. In my vision, my feet were the foundation of the material world. As I inhaled, the material density of my body seemed to dissolve and I "popped" into the bright world of living geometry and infinite spirit. I noticed strange jewel-like chakra centers within my glowing wire-frame spirit body and spectral colors that were absent from my dream painting. I was *in* my future painting and was being given an experience of the state in order to better create it.

After receiving these two visionary encounters of the same painting, I began to draw what I had seen in my sketchbook. This started the translation phase, bringing the inner solution of my artistic problem to an outward form. I drew the body and worked on the computer to plot an accurate texture map of the electric grid around the hypermind sphere.

THE ESSENTIAL INSIGHT

The teaching provided by each of these experiences is somewhat similar, and it relates to the Akashic field. There is a Universal Creative Force connecting all beings and things, a source of love and wisdom that can be drawn from and revealed through the creation of works of art. For myself, as well as other artists, entheogens have played a crucial role in the visionary connection to the Universal Mind of the Akashic field in the creative process. However, I don't advocate that artists live in a constant haze of chemically altered consciousness, and some sensitive artists should completely steer clear of the substances. Meditation is the royal road and it has been extremely helpful for me in catalyzing visionary mystical dimensions of consciousness.

However one connects with the Universal Mind and the Akashic field, may the subtle inner worlds of imagination and illumination open an endless source of inspiration for a new and universal sacred art.

THIRTEEN

Reconnecting to the Field

Eric Pearl

Eric Pearl has trained over 45,000 people in more than sixty countries in Reconnective Healing. He is the author of *The Reconnection: Heal Others, Heal Yourself,* published in more than thirty languages. He has presented at the United Nations and Madison Square Garden, and his seminars have been featured in the *New York Times*. Pearl lectures and gives training seminars worldwide.

I want you to wonder . . .
I want you to realize another plane . . .
Part of a much greater plane . . .

Let me say first of all that when Ervin Laszlo asked me to write a contribution for his book on the Akashic experience—the experience of communication with the Akashic field—I had no idea where to begin. I was so focused on telling just the *right* story for this book that I didn't see the forest for the trees. But here now is *my* story. It can also be *your* story: all of us can access the field.

My first big "aha!" experience—at least the first that I may have been willing to recognize—came when I was suddenly awakened in the middle of the night by a bright light that came piercing through my eyelids. I opened my eyes to find that it wasn't anything too seemingly spiritual at all. It was simply the lamp next to my bed. It had decided to turn itself on.

As my logical mind was telling me that it must have been some kind of an electrical "short," or *reverse* short if there were such a thing, another part of me that wasn't too interested in my exercise in logic became acutely aware that there was someone in my house. I can't tell you what an unsettling feeling it is to awaken to someone being in your house who wasn't there when you went to sleep. Let's just say I got up with a knife, a can of pepper spray, and my Doberman pinscher . . . and went looking. Eventually it became clear that there was no one there that I would discover that night, at least in a form that I consciously anticipated. I went back to sleep, still not completely certain that there wasn't someone hiding in my home, somewhere.

The following Monday, I went into my office to see my patients, just as I would do on pretty much any Monday. I gave my first patient a chiropractic adjustment and then told him to close his eyes, relax, and allow the adjustment to settle into place for a minute or so. When he opened his eyes, he asked me who had come into the room while his eyes were closed. I explained that no one had been in. He said, "No," as if I hadn't heard the question correctly, "Who came into the room while I was lying here with my eyes closed?"

"No one," I responded.

He continued, "The person who stood in the doorway."

I said, "No one stood in the doorway."

"I heard them come in," he practically pleaded.

"No one came in."

"I felt them standing there," his tone became somewhat angry, as if I weren't telling him the whole truth.

We went back and forth and back and forth until finally he said, "Okay." But the look on his face told me that he didn't quite believe me.

I thought nothing of this outside of the fact that it was a little out of character for this person and went into another room to see my next patient. When we finished, she opened her eyes and asked, "Who walked into the room while I was lying here on the table?"

"No one," I replied, finding this simultaneously amusing and eerie.

She went on, "I heard them come in."

"No one came in."

"I felt them there," she insisted.

All I can tell you is that on this one day, seven of my patients independently and adamantly informed me that while they lay on the table, someone stood in the doorway, came into the room, and walked or ran around the room. Two of my patients actually looked me straight in the eye, voices somewhat lowered, and said it felt as if someone was *flying around the ceiling.*

This was my twelfth year in practice. No one had ever said anything like this to me before. And yet on this one day, seven of my patients, with total personal surety, informed me of their strikingly similar experiences. Strikingly similar to one another—and strikingly similar to what I had experienced just a few nights earlier.

As if this were not enough, other patients on the very same day told me that they could feel my hands before I touched them.

"Oh, sure you can," I said, disbelievingly. "Close your eyes."

As their eyes were closed, I'd move my hands over different parts of their bodies, my palm aimed toward them from distances of several feet to a few yards or more away. It seemed that the farther away I'd move, the stronger their sensations and the more acute their awareness, just the opposite of what's experienced with energy healing.

"Left shoulder," they'd say. "Right ankle."

They knew. Somehow they knew right where I was holding my hand, or right where I was directing it. Sometimes they felt heat, yet more often they reported feeling a definitive "cool breeze." As I continued to play this game with my patients, unexpected physiological occurrences began to show themselves. My palm blistered on more than one occasion. And then it bled. Not like a stigmata, more like the way it might bleed if it had accidentally been stuck with a pin—with that little drop of blood that doesn't really go anywhere. Yet suddenly people began to have healings. Unanticipated healings. People were getting up out of wheelchairs, reporting that their cancers had vanished while brandishing laboratory reports, X-rays, CAT scans, and whatever else they could find in case I didn't believe them. Mothers reported that their epileptic children stopped having seizures, and cerebral palsy children regained the ability to walk and speak. Not all of them. But a lot of them. Enough of them that we knew something real was going on.

When my patients would ask what I'd done, I responded, "Nothing. And don't tell anyone." Soon people started coming in from everywhere saying, "I'll have what she had." "Teach it," they said.

"Teach it?" "How the heck can I teach it?" I thought. "No one left an instruction manual at my feet. All I do is wave my hands in the air."

About all I could think of saying was, "Wave your hands in the air." But I knew there really was more to it. I knew what I felt. I knew how I *found* then *followed* what I was feeling and learned how to recognize it with ever-increasing clarity. Could I explain it? And how could I teach it? Wouldn't I have to *understand* it first? And should I allow my lack of consciously possessing a full understanding of it to interfere with my sharing the insights that I *did* have and that were constantly unfolding and revealing themselves to me? In other words, could I teach this as a work in progress? More importantly, *could I not*?

Patients would phone and tell me that after their sessions, when they walked into their houses, a lamp or television would turn itself on and off repeatedly and they'd experience strange sensations in their hands. They went on to say things like they had instinctively held their hands near someone in their family, and their grandfather regained his ability to speak after a stroke, or their aunt regained the use of her arms . . . None of these things, however, had been a part of their daily life prior to that moment and, for most people, none of these things had ever even crossed their mind.

BEGINNING RECONNECTIVE HEALING

I agreed to teach a class, it was of all of approximately twenty-five people. Not having what is classically considered an organized or disciplined nature, I found myself making my first attempt at creating notes and an outline for the class while I was driving there, late and lost. I walked into the room, looked at the notes, and knew that any attempt to follow them would be futile: foolish at best, and more likely an awkward distraction. I opened a massage table, sat down on it, and told the story of how this had come about and unfolded to that date as best I could.

Then I let everyone there feel in their hands the sensations of what is recognized worldwide today as a new level of healing called *Reconnective Healing*. I had them take turns lying down on the table to experience it and standing up to learn how to work with it. At ten minutes prior to the designated finish time for the class, I suddenly had nothing left to say (an affliction from which I no longer suffer) and ended the evening. I then sent a group of twenty-five new Reconnective Healing practitioners out onto an unsuspecting planet and told them to call me if anything interesting happened.

They phoned. And phoned. The phones rang so much that I had to hire another person in the office just to answer them and purchased my first computer to respond to all the inquiries. The inquiries for healing sessions. And the inquiries to teach. And eventually I wrote a book called *The Reconnection: Heal Others, Heal Yourself* so that you can discover how to access this more fully—to help yourself and to help others.

In my practice, as well as in the seminars I began to teach, I continued to discover just how nonlocal *nonlocal* can be. I saw that the hands are used in this expanded level of healing essentially to allow people a feeling and a familiarity, and to have the experience themselves. Once they do, I show them how to do this from meters away from the person they're working with, which is what I was doing when people first noticed that they were "feeling" my hands touch them. Some of the time

I was a few feet or yards away, other times across the room or even out in the hallway, watching.

As the seminars advanced in size and attendance, we had to employ large video screens so that the demonstrations could be seen. I would often bring some people up from the audience who had neither read *The Reconnection* nor practiced any form of what's classically considered "energy healing." I would show them how to *feel* this comprehensive spectrum in their hands, then allow one of them to work with it on stage to show the audience that they can instantly bring about the same effects that I do, effects that are consistently reproduced in controlled, randomized, double-blind studies. Then I have that person begin to walk away, *pulling* the sensation along with them, and all the while, on-screen, we clearly watch the involuntary physical responses of the person lying on the massage table as they respond to the Reconnective Healing spectrum. These "registers" become stronger and more clearly evident as the person who is standing walks farther and farther away.

During one of these times I became daring and had the person doing the work watch the screen instead of watching the person on the massage table, with the screen turned at an angle so that he couldn't see the person on the table at all. That proved just as effective. As the seminars became even larger we found that we had to occupy multiple rooms in a convention center or other venue to accommodate everyone. I would take the person who was doing the work outside the main room to a different location and have him or her work from there on the person who was on the stage in the main room. The effects were just as clear and vivid, and they proved fairly consistently stronger than when we were in the same room.

Today, toward the end of the seminars, I show people how they can do this with their eyes. Everyone can do it *once they interact with it*—including people who were blind from birth. Not just legally blind, but *blind*. One woman didn't even have her eyes, just prosthetics. It's not the physical eyes. Is it the third eye? *Is* there a third eye?

The experience of having someone feel my hands and the hands of other Reconnective Healing practitioners on their bodies might, at first glance, be considered a local experience, since it *could* be transmitted by the body's biofield. But when the same effect is transmitted *beyond* the likely range of the biofield, then it's a nonlocal effect. Is the healing local when two people are in the same room? And is it nonlocal when a wall separates them? How far away must that wall be? Does a video screen or monitor in a different room make it more nonlocal? Does healing with the eyes?

BRINGING LIGHT AND INFORMATION

Let me move now to a part of this report that gets still stranger. One January day, a patient of mine lost consciousness—or *gained* consciousness, depending on your perspective. In this state he spoke—or, more accurately, *someone* spoke *through* him—two distinct phrases: "We are here to tell you to continue doing what you are doing." "What you are doing is bringing *light* and *information* onto the planet." Two days later, it happened to three other patients. Three days later, it happened to five. Over a three-month period—the second week in January through maybe the third week in April of that year—over fifty of my patients had this same experience and spoke a total of six verbatim phrases. None of them, except for the first patient, ever had had a voice other than their own come through them, and none of them knew any of the other patients this happened to.

"Continue doing what you are doing." What could that mean? Well, for starters, what it *doesn't* mean is to simply do the same thing over and over again. *Continue* isn't a word of stagnation, it's a word of movement. Continue . . . continue forward . . . continuation . . . *continuum.*

The second phrase, "What you are doing is bringing light and information onto the planet," was equally fascinating. Why *light and information*? Why not *light and love*? I received some insight into that question from people who've had a "life-after-death" experience. I know it's usual to refer to this as a near-death experience; I, however, prefer to listen to what those who have had the experience say. And what they say is that they weren't *near* death, they *died*. And what they say is that the light that they see when they're dead *is* love, and the love is indescribable. Most likely we know inside of us that there's more to light than what the word *light* conveys and that there's more to love than what the word *love* conveys. Therefore our desire to express something that we know is more than something we have words for results in the well-intentioned, unconscious, new age linguistic redundancy of "light *and* love."

This still doesn't explain why the words *light and information* were selected. Many who have looked into this feel that at least one potential insight into this question can be found by looking at the photon, known for its properties of energy, light, and information. So why didn't the phrase come through as "What you are doing is bringing *energy,* light, and information onto the planet"? Because, as best as we can tell, the energy was already here; it was what we had already been accessing. Light and information are what we are able to access to expand beyond energy and energy healing and to transcend its myriad techniques and offshoots—old and new.

Let's think of it like this. If light is consciousness, with expanded light comes expanded consciousness. With an expanded consciousness we have more information available to us. We can evolve beyond the need to move our hands in one direction instead of another, the need to sweep up in the front or down in the back (and allow ourselves to see past the illusion of up/down/front/back altogether), to protect ourselves in white, violet, or gold flames or by wearing the specially encoded necklace du jour, to inhale to one count and exhale to another, to touch specific points, intend wavelength collapses, spin chakras clockwise or counterclockwise, or to blow out, shake off, or spray away "negative" energies. An expanded consciousness allows us to see past "negative" energy—as it has all too often been viewed in the healing world—and to recognize that darkness doesn't exist, for if it did, we could sweep it up, place it into a paper bag, and put it out with the trash. We recognize instead today that it's only a place where we have not yet allowed ourselves to shine as the light. The fear-based concepts of "evil" and "negative" ("bad") energy in healing, and in the field itself, turn out to

be illusion. The gift is the demystification. The challenge is our willingness to allow it to be demystified.

HEALING WITH THE FIELD

How do you work with the field? How are people who leave a Reconnective Healing seminar able to access more clearly and more comprehensively a multidimensional field than they were prior to their arrival, or than they were through the various energy healing techniques they previously had? And how can this be taught in a way that allows people to get it?

It isn't really possible to *teach* healing or field interaction. Field interaction and healing happen to some degree *anyway*. It's a gift, a part of our existence, our presence, our natural function. And yet there's more. And the *more* comes about simply when we allow ourselves to open our awareness to it. It comes about once we see it, feel it, and allow ourselves to recognize what we see and feel, and once we give ourselves permission to *acknowledge it*. For without acknowledgment, our ability to access it fades, because our willingness to access it fades.

The *more* comes about once we allow ourselves to listen . . . *with a different sense*. To stop *doing* and, instead, *to become:* to become the observer and the observed. This is where and when the universe often chooses to display its wonder and its beauty for us. This is the gift. We then see things that are new, that are different, that are real. Very, very real. We experience, then, each session with a sense of newness and discovery because, with each person, it *is* new. A doorway opened for us is one that, in turn, allows us to open doors for others. This door is a more conscious and aware interaction with the field, and more rewarding for all involved. It doesn't feed the ego of the doer; instead it gives higher and lasting sustenance to all involved with unparalleled healing results. The nature and form of the healing is determined not by the limited, conscious, educated human mind and the ego of doing, but rather by the intelligence of the field, the intelligence of the universe, zero point field, the Holy Spirit, God, Love, spirituality—or *the Akashic field*.

Once we get out of our own way and allow ourselves to access the Akashic field of spirituality, healing, intelligence, and evolution, we allow ourselves to experience *oneness*. And once we truly experience it, we find out just how normal that really is.

This is healing of a very different nature than we have been taught to perceive, understand, or even believe or accept. This healing is about an evolutionary process brought into existence through *cocreation* at the highest level of interaction with the universe, at the highest level of interaction with the field.

We are all here as one.
We have our own vibration . . . but we are of the same sound.
We all have the music of life.

FOURTEEN

Shaping Creative Fields

LESSONS FROM MY AKASHIC EXPERIENCES

Masami Saionji

Masami Saionji is a native of Tokyo, Japan, and serves as the chairperson of the Goi Peace Foundation and the World Peace Prayer Society. A descendant of the royal Ryukyu family of Okinawa, she continues the work of her adoptive father, Masahisa Goi, who initiated a movement for world peace. Saionji is the author of twenty books and together with her husband, Hiroo Saionji, was awarded the Philosopher Saint Shree Dnyaneshwara World Peace Prize in 2008. She is a member of the Club of Budapest and the World Wisdom Council.

There is, indeed, an Akashic field containing information on all that has occurred since the proverbial "big bang." This information spans the birth of galaxies and of tiny one-celled creatures. It holds a memory of each thought, word, or movement that has been generated since the beginning of time. To be sure, these Akashic memories include a record not only of what occurred on tangible, visible planes but also on intangible, invisible planes as well. It is fitting, therefore, that Ervin Laszlo should dedicate an entire volume to study the Akashic experience: the varieties of experiences that give us access to this all-inclusive Akashic field.

A MYSTICAL EXPERIENCE

My earliest memory of a mystical Akashic experience is from the time when I was nineteen years old. I suddenly became ill during a family trip to Okinawa. I fell unconscious at a site where, at the end of World War II, many young Okinawan girls had committed suicide. I was rushed to the hospital and later taken back to Tokyo.

My illness was diagnosed as a cerebral tumor, and the option of surgery was deemed unpromising. For the next year I had seizures every day and was unable to keep food down. I became very thin and gradually lost my eyesight, then my hearing.

Starting from the early days of my illness, I was visited every day by Goi Sensei, whom I had met at the age of fifteen, before I started volunteering in his World Peace Prayer movement.[1] After I became ill, Goi Sensei would pray with me each day, healing me with purifying energy.

Goi Sensei had often told me about my guardian angels (he called them "guardian spirits and guardian deities"). However, until then, I had always refused to listen because I disliked topics relating to anything psychic or spiritual. While I was ill, however, my feelings changed and I got into the habit of praying to my guardian deity.

During my illness, I used to see grotesque visions of human beings without faces or with missing eyes, noses, or other bodily features. But thanks to Goi Sensei's encouragement, I never gave in to fear or tried to run away from these phantom beings. I steadily continued to pray and spoke to them reassuringly, saying things like: "You are all right! God loves you. God is taking care of you. You will be fine! Give thanks to God. Think of peace. Think 'May peace prevail on Earth.'"

In time, these visions gradually diminished and ceased. Gradually, my hearing returned, as did my eyesight. But for a long time I was weak and continued to have seizures. Feeling that my existence must surely be a burden on my family, I often thought about dying.

WHEN I LOOKED AT THE SUN . . .

Whenever I felt fearful or uneasy, I would pray. One day, while I was praying intently, I happened to glance up at the window above my bed. It was about midday. Since my eyesight was getting better, I was able to see the shining sun. I remember thinking: "Oh, the sun is shining! How lovely and warm it is. It melts the cold and warms my body. Thank you, dear sun! Thank you!"

As I was thinking this, the sun began to approach me. It kept getting bigger and bigger. I felt as though it was falling straight down from the sky. The sun came closer and closer and, when it was right in front of me, it merged into my body.

Just then, I saw the light of my guardian deity. It was a warm, loving light—not a human form—and I knew at once that this was my guardian deity. She spoke to me, not with words, but with an instantaneous flash of meaning lasting for less than a second. Put into words, this is what she said:

> *I am always with you, just as you see me now. I am your guardian deity. Everyone has a guardian deity, shining directly behind them at all times, guiding and protecting them. I know how much you have been suffering, but you must stay in this world. You have a mission here. Your mission is to tell others that guardian deities and guardian spirits really do exist. Now that you have seen me with your own eyes, you know that it is true, so you can surely tell them. If you don't do anything else from now on it is all right. Just tell people about their guardian spirits and guardian deities.*

When I received this flash of meaning from my guardian deity, I felt gently enfolded in her love. I felt totally comfortable and at peace. It was then that I noticed that I was no longer in my physical body. I was looking down at my body from above. I could see the energy flowing out from my fingers and toes. "Oh," I thought, "the energy is leaving my body. How interesting it is. This must be what death is!" Then, recalling what my guardian deity had said, I was instantly back in my body, lying in my bed.

A MARVELOUS INFLUX OF UNIVERSAL ENERGY

It was after this near-death experience that I gained a deep sense of the infiniteness of the universe. I became clearly aware that there is indeed a wellspring of life—an expanse of unlimited potential—existing at the source of the universe. From one moment to the next, each of us is sustained by the life energy that it is continually emitting to us.

After the big bang, the wellspring of the universe radiated various fields of life—mineral life, plant life, animal life, human life, and so on. Human life was imbued with a creative capacity—the ability to generate creative fields. Whether or not we intend to, we are constantly creating. For human beings, to exist means to create.

When human beings became confused and started creating unharmonious conditions, the wellspring of the universe radiated new fields of divine life whose purpose was to provide love and guidance to human beings. These fields can be described as "guardian angels," or "guardian spirits and deities." Ever since my near-death experience I have been constantly aware of the presence of these protective beings.

Even now, the wellspring of the universe is always sending energy to human beings. It does not send one kind of energy to one person and a different kind to another. It sends the same kind of energy to everyone. But each person's way of using this energy is different. How we use this energy determines the course of our future and the future of the world.

At any moment, a human being can receive new energy from the wellspring of the universe. However, to receive this pristine energy, our thoughts need to connect with it. We can do this with pure prayer or with peace-loving thoughts such as "I am so grateful! Everything will definitely get better!" or "May peace prevail on Earth!"

Creative fields that are bright and positive in nature are able to connect with the harmonious universal intention. So, no matter how many destructive fields might exist in this world, we must never feel discouraged. When we use our energy for positive aims—aims that are good for Earth and for all things—we will be supported by a marvelous influx of universal energy. At that time, the creative fields for world peace will surely expand and grow. This process has already started.

MY MEDITATIVE AKASHIC EXPERIENCE

When I go into meditation, I simply concentrate on oneness with the essential universal vibration. My breathing becomes deeper, slower, and more spiritualized. My body cells become more spiritualized as well. I can feel my cells expanding through the surrounding space until there is no dividing line between me and space, nor between me and others. My consciousness still recognizes that I am Masami Saionji, but there is no borderline to my existence. I am one with the universe, one with Earth, one with all living things, existing beyond time and space.

Because we are all connected, we can understand each other's mind and feelings. We can understand what happened in the past and how to guide the future. When we have experienced this oneness, there is no further need for us to strive, or endure, or make efforts—no need to learn from any teacher. We only need to be in oneness. We only need to make the connection.

Until we make this connection, we need to keep turning our thoughts to the wellspring of life, the source of infinite potential. We must not give up. Even if the goal seems far away, we must take one step forward. We must try, try, and try.

We must always keep in mind that, in truth, there are no dividing lines between people. When we know that dividing lines don't exist, we understand that when we speak, think, or move, the effects instantly spread to others. This knowledge brings us a sense of responsibility. We know that we must exert a positive influence on others.

And so, we need to brighten our consciousness, bringing it closer and closer to the universal source. That brightness will instantly touch others, sparking them to evolve and create in a positive direction.

If we want to avoid giving our energy to unharmonious creative fields, we need to closely observe, and take positive control of, our words, thoughts, and actions. In this way, unharmonious creative fields will eventually disappear, and only bright ones will remain.

CREATIVE FIELDS

Starting from the age of nineteen or twenty, after I had my near-death experience and met my guardian deity, mystical and meditative Akashic experiences gradually became an integral part of my life. While living in this physical realm, and interacting with people from day to day, I am at all times clearly aware of the spirituality of each of them. I naturally hear the voices of plants, animals, and other living things, and I feel their joys and sorrows as keenly as if they were my own.

When I look at the people around me, I see the same attributes in them that are visible to anyone else: their physical form, the clothes they are wearing, their posture, their gestures, and their facial expressions. At the same time, I also see the fields of creative energy that they are generating.

The thoughts, words, and emotions that human beings emit from moment to moment are constantly streaming forth from their bodies, forming creative fields that are visible in various colors, forms, and shapes. Some of these creative fields are hazy and cloudy in appearance, and they hover around the person in indeterminate patterns. Others are extremely tenacious, and they twine tightly around the person like a resilient string.

Emotions like happiness and affection appear in shining colors, and they envelop the person in a halo-like radiance. Emotions like worry, fear, and animosity reflect smoky colors, and they adhere closely to the person in a suffocating manner.

Each thought or emotion holds its own unique form of creative energy, and thoughts having a similar frequency band together to form a homogeneous creative field around the person who emitted them. When the amassed energy of a particular creative field has accumulated to a critical point, and is triggered by some external circumstance, it manifests itself in some way on the visible plane. This manifestation might occur in the form of some sort of event, happening, or situation that emerges in the person's life. It might take shape in the form of an encounter with another person. Or, it might manifest itself in words that are heard or read, or observed popping up unexpectedly in the person's own mind. Once this manifestation has occurred, the energy held within the creative field is depleted to just that extent.

THE MANY KINDS OF CREATIVE FIELDS

From what I have observed, there are as many kinds of creative fields as there are human thoughts and emotions. On the positive side, there are creative fields for qualities such as thoughtfulness, optimism, gentleness, dignity, purity, cleanliness, goodness, and sincerity. There are creative fields for feelings like happiness, enthusiasm, friendship, courage, admiration, gratitude, trust, respect for life, respect for nature, and humanitarian love. There are creative fields for behavior such as praise, encouragement, acceptance, generosity, and forgiveness. There are creative fields for phenomena like inspiration, healing, fulfillment, improvement, and accomplishment. And there are creative fields for conditions such as peace, harmony, truth, sanctity, abundance, and bliss.

On the negative side, there are creative fields for feelings like pessimism, frustration, anxiety, self-blame, self-pity, self-doubt, self-hate, self-justification, guilt, and a thirst for revenge. There are creative fields for phenomena such as discrimination, suicide, accidents, wars, and disasters. There are also creative fields for fixed ideas, such as a belief in sin and punishment, a belief in defeat, a belief in failure, and a belief in illness and poverty.

Simply by reading these words, we can gain a sense of the emotional energy that sustains them. Words themselves are a concentrated form of energy and a powerful means of creation.

HOW DO CREATIVE FIELDS GET STARTED?

Because creative fields like these give form to our personalities, our health, our relationships, and the conditions that surround us, I would like to discuss briefly how they get started.

Where does our creative energy come from in the first place? Pure energy is always emitting from the source of the universe, brimming with infinite potential. This energy supports all forms of life, big or small. Each human being receives a steady supply of this same universal energy, and each of us assigns a shape to that energy by means of our words, thoughts, and emotions. This is the process of creation.

The energy of any thought, whatever it might be, holds the potential for generating a creative field. However, unless several thoughts of the same type are emitted, a creative field will not be formed. Just one thought, in and of itself, will quickly lose impetus unless it is reinforced by the energy of other thoughts that resemble it. If several thoughts of the same type are emitted, the energy of those thoughts will band together and the rudiments of a creative field will take shape. Once this rudimentary field has been formed, new energy can easily be added to it. As its mass increases, it will consolidate into a cohesive creative field. As this field continues to grow, it exerts a stronger and stronger influence on the person's will, decisions, and behavior.

COLLECTIVE CREATIVE FIELDS

The activity of a creative field becomes further invigorated when similar creative fields are generated by other people. This is because creative fields having the same vibrational frequency tend to merge, forming large-scale collective creative fields. These collective creative fields can be extremely pervasive. The bigger they are, the more powerfully they respond to thoughts emanating from people in various places.

Let us say, for example, that someone idly thinks of committing suicide. To begin with, his wish to commit suicide is not particularly strong. Unfortunately, however, his idle thoughts attract the vibrations of a big collective creative field for committing suicide. These vibrations enter his body and exert considerable influence on his thought processes and behavior. Unless positive influences intervene, or unless the person consciously rejects those negative vibrations, he or she might end up actually committing suicide.

When it comes to positive thoughts, the scenario becomes much brighter. Here is an example of a real-life experience of one of my readers.

For a long time, this person had been in the habit of underestimating herself. Even over trivial things, she would find fault with herself and think uncomplimentary words such as: "I am selfish." "My soul is tarnished." "I am unworthy." "I am unappreciated." "I am unloved."

Each time such words passed through her mind, they would summon the arrival of similar types of energy from large-scale collective creative fields. She would then be invaded by an onrush of self-deprecating energy. The influx of this energy would then plunge her into a state of intense unhappiness. As a result, she found it quite difficult to stay in a good mood from day to day.

After coming in touch with my theory of *effect-and-cause,*[2] she realized what a disservice she had been doing to herself with her own thinking. With firm determination, she made up her mind to create an entirely new way of living based on positive words. She chose two words that were particularly attractive to her: "purified" and "spiritualized." She then created a kind of litany around them, combining the words with her own name. It went like this (for the purpose of this example I will call her "Annie Hunt"):

> *Purified Annie Hunt, spiritualized Annie Hunt.*
> *Purified Annie Hunt, spiritualized Annie Hunt.*
> *Purified Annie Hunt, spiritualized Annie Hunt.*
> *How can we thank you? May peace prevail on Earth.*
> *On behalf of humanity, we thank the love of the universe*
> *for the purified existence of Annie Hunt.*

Again and again she would repeat these words to herself in a continuous rhythm—like a poem or a melody. When disconcerting emotions arose, she would inhale deeply, hold her breath, and affirm the first few words in her mind. Later on, she would vary the words to suit the moment. To uplift her thoughts regarding other people, she also tried creating similar litanies using other people's names.

Because to respect one's own existence, and the existence of others, is in tune with the great universal will, these positive words attracted a flow of shining energy from all over the universe. And because she added the humanitarian words "May peace prevail on Earth," as well as words of gratitude for universal love, she was able to access the incredibly uplifting power of a higher-dimensional, universal energy. Not only did this bring happiness to herself, it also helped establish a more cohesive collective field for happiness on planet Earth.

TAKING CONTROL OF OUR CREATIVE PROCESSES

Much has been written and said about the creative power of thoughts and words—especially words. But it seems to me that few people truly understand this power. If they understood it, how could they go on using words with such carelessness? Just like children playing with fire, human beings are spewing out negative words without seeming to care what the effects might be. Here I would like to clearly declare that the energy of our every word, once it has been uttered, flies into the creative fields that are forming around us, greatly intensifying their activity. Not only do these creative fields generate our individual happiness or unhappiness, but they also give rise to world conditions of poverty or abundance, respect or discrimination, environmental destruction or rebirth, war or peace.

How can we take control of our creative processes? How can we dissolve the unharmonious fields that we have unknowingly generated? Our first step, I feel, should be to steadily monitor our own thoughts and words to gain a sense of what kinds of fields we are creating. Next,

I suggest that we train ourselves to cancel all negative words as soon as they come to mind by replacing them with positive ones. At the same time, I recommend that we make up our minds to banish our destructive beliefs by transforming them into constructive ones.

To do this, we must clearly know that our energy flows in whatever direction we are thinking. If we think of and believe in our potential for building a peaceful, happy world, our thought-energy jumps ahead of us and forms a creative field for that situation. As that field fills with energy, its power increases. Eventually, it becomes so strong that it can pull us toward the peaceful future that we ourselves have envisioned. When we have reached that stage, it is no longer necessary for us to make vigorous efforts. We are so closely connected with the creative field for world peace that we can merge into it without effort.

It might be effective for us to envision small-scale goals, and perhaps most of us are doing this naturally. I feel, however, that to effect large-scale changes in our lives and in the world, we need to hold large-scale visions in our minds and consciously pour our energy into them.

DEACTIVATING DESTRUCTIVE FIELDS

An increasing number of people are deactivating destructive fields by means of positive affirmations. Like the person mentioned earlier (whom I called "Annie Hunt"), each of us can use the words of our choice. Likewise, our preferred affirmations can change with our changing circumstances. An affirmation can be a prayer, a poem, a belief, or a goal. The important thing is to choose an affirmation that's meaningful for us and that in some way contributes to the happiness of humanity and the whole Earth.[3] I suggest that each person try various affirmations to see which ones match his or her aims and personality.

How is a positive affirmation practiced? Let us say that you have chosen the positive words "Everything will definitely get better!" Whenever you have a chance, say this phrase aloud over and over again. Or, you could try inhaling deeply, then repeating the phrase in your mind many times while holding your breath. Many people have found that, through this practice, their thought habits really do change, and their circumstances take a turn for the better.

Another example of a positive affirmation is "I am so grateful!" Many people find that by engraving these words into their consciousness, they naturally begin to feel grateful toward Earth, their family, their friends, and everything in creation. In consciously declaring that they are grateful, they naturally give birth to their own creative field for gratitude. This, in turn, merges with other such creative fields, making it easier for many people to experience grateful feelings. When the spirit of gratitude enfolds Earth, nature will be revived and all living things—ourselves included—will be able to live together in harmony.

When I observe people practicing affirmations of gratitude, or praying intently for peace on Earth, I see a clear white light emanating from their bodies. This light envelops the destructive fields that project wars, environmental disasters, and other tragedies, softening them and reducing the scale of their activity.

Each of us is responsible for building positive creative fields around us, using our own spiritual and physical energy. Each human being is responsible for connecting with mystical universal energy and drawing it toward us from the future.

The day is not far off, I believe, when each human being will be able to access the memories contained in the Akashic Record. There we will find tales of civilizations that have bloomed and perished. We will observe how they met with destruction after intruding into sacred fields and violating the laws of nature and truth. Present-day humanity is now headed in that direction. However, by following the example of awakened forerunners, we can make a positive shift.

If humanity is to take a better direction, each of us must gain our inspiration not from the past, but from the future. We must continue to focus on this thought. We must try, try, and try. We must never give up.

In the Akashic field, wonderfully bright new information is waiting to be recorded. We are the ones who must create it. *May Peace Prevail on Earth.*

Part Three

RESEARCHING THE EXPERIENCE

FIFTEEN

Exploring the Akashic Experience

BRIDGING SUBJECTIVE AND OBJECTIVE WAYS OF KNOWING

Marilyn Mandala Schlitz

Marilyn Mandala Schlitz, anthropologist, has pioneered laboratory, clinical, and field-based research in the area of consciousness, culture, and healing for three decades. A researcher, speaker, change consultant, and writer, she is the author of four books and more than 200 studies and articles. Schlitz is vice president for research and education at the Institute of Noetic Sciences, senior scientist at California Pacific Medical Center, and chief learning officer of Integral Health Network.

Where does one's story start? I begin mine with what I don't remember.

At eighteen months old, as I am told by my family, I found a can of lighter fluid on the table. Being a curious child, I did what curious children do: I put it into my mouth. For months after, my small body rested and wrestled in a hospital, floating in and out of life as my lungs sought the affirmation of breath.

Perhaps it was here, in the entrusted hands of a dedicated group of health professionals, in the prayers and intentions of my devoted family, in a personal biological quest for life itself, even under dismal odds, that I developed my abiding fascination with healing.

As I trace my history, I am aware of the various seeds and fertilizers that have led me to a career far outside the mainstream, one in which I have sought to understand the interface of science and the Akashic field. It was a series of small, exceptional experiences that paved the way for both my personal transformation and my professional contributions to an emerging new worldview.

HEALING AND TRANSFORMATION

I grew up in Detroit, Michigan, in the 1960s and '70s. This was a time when the United States was at war with itself. It was a war of race, of class, and ultimately of what I came to see as a war of consciousness and worldview.

Coming of age in such a complex time and in a setting that fueled rebellion at individual and social levels, I was alive with confusion, anger, and a desire for change. One night, when I was fifteen, I was with someone I should not have been with, doing something I should not have been doing. A drunk driver pulled out of a parking lot of a bar without his lights on and hit the motorcycle I was a passenger on, throwing my body through the air. I clearly recall watching my physical being tumble through the sky and then crash to the ground.

During what I now understand as an out-of-body experience, I felt my awareness transcend my body, looking down on it from a higher vantage point. It was an opening to some capacity that didn't have a frame of reference or a language in my limited life experience. During an extended time in the emergency room, waiting for my parents, who were hours away, there was talk of possible amputation. The cut was deep and wide in my left leg. The emergency medical team did their best, putting sixty-six stitches below my knee and sending me home with question marks about my recovery.

Later, lying on the couch in my family's home, I somehow got the idea that I could and should visualize my immune system healing my leg. I lay there for long periods of time, feeling the tingles of healing. I didn't come from a medical family, and I have no memory of having heard about mind-body medicine in the early 1970s. Now I can see that I had a direct, noetic understanding about what I needed to do to bring about my own healing. Today I have two well-positioned feet on the ground and an awareness of some aspects of myself that are more than just the physical.

PSYCHIC EXPLORATION

When I was an undergraduate, I had several meaningful events that helped shape my life. The first was reading Thomas Kuhn's book *The Structure of Scientific Revolution*.[1] This book, and the idea that our paradigms of reality are socially constructed and not absolute, was nothing short of a conceptual liberation. It gave me hope that the failing vision of society that was around me need not be final or binding. Indeed, even in the context of science, we have experienced different models of reality—one replacing another. What was needed for our society, I felt sure, was a fundamental, whole systems transformation.

Piggybacking on this insight came many conversations with a professor of neuroscience, Robin Baracco. During lengthy discussions, many years before the neuroscience revolution, I learned about how much we knew—and still don't know about consciousness, brain, and mind. One day Dr. Baracco gave me another book, *Psychic Exploration* by Edgar Mitchell, the Apollo 14 astronaut.[2] This volume completely changed my life and sent me on an intellectual odyssey that compels me to this day. The idea that a serious group of scientists was exploring psychic or psi phenomena seemed to me to be the harbinger of a new paradigm I felt sure we needed. Instead of placing my focus on the material aspects of reality, endemic to the dominant culture, I decided to commit myself to understanding the transformative potentials of human consciousness.

I began doing very preliminary remote-viewing experiments with experimental psychologist Charles Solley, at Wayne State University. We spent the summer of 1977 personally testing the claims made by physicists Russell Targ and Harold Puthoff in their book, *Mind-Reach*, that people can describe geographical locations about which they have no sensory information.[3] We were our own subjects and experimenters; the results were startling and compelling.

On the first day, we invited a self-proclaimed psychic to visit the psychology lab and be our subject. I was the "outbound" experimenter in the first session, the one who visited a distant site. The description given by the psychic didn't even come close to matching the geographical site that I viewed. We tried a second trial. This time I served as the "inbound" experimenter, sitting with the subject in the lab and asking for her impressions of the location of Dr. Solley, the outbound experimenter.

When Dr. Solley returned, he asked the psychic to describe her impressions. Again, she described almost nothing that matched the site. Remembering that he and I had both produced strong matches in our initial explorations, he then asked me if I'd had any impressions. I was quick to say no; I was the objective experimenter, after all. He pushed a little; surely I had some impressions? I conceded that a strange symbol (resembling the Greek letter *omega*) had come to my mind, and I made a quick drawing. He grew excited and took us to the building where he had been sitting during the sending period. Sure enough, there was a fence surrounding the building that was made up of the symbol I had drawn, and etched on the side of the building was the symbol. This was my first insight into the potential myth of objectivity in our studies of consciousness. It was also one of the first times I gained some direct, first-person experience of the phenomena I was hoping to study.

EXPLORING PSI IN THE LAB

Since this early phase of my career, more than three decades ago, I have had many compelling encounters with the Akashic experience. Unlike most people, however, many of my personal encounters have occurred in the context of well-controlled laboratory experiments. Let me consider three specific examples from my formal research.

Remote Viewing

In 1980 I conducted a remote-viewing experiment with Elmar Gruber in which we were both the subjects and the experimenters.[4] We designed a formal study of ten trials over thousands of kilometers between Detroit, Michigan, and Rome, Italy. Gruber selected a pool of geographical target sites around Rome that were unknown to me. On each of ten experimental days, the outbound experimenter randomly selected and then visited one site. At the same time, I sat down to describe his physical location without any sensory knowledge of where he was at that time. During the experimental periods, we had two colleagues keep copies of our list of targets and the drawings for each day; this was done for security reasons. Following completion of the ten trials, the ten descriptions were randomly presented to five independent analysts. Each was asked to independently visit the different locations and to evaluate the degree to which the site matched each of the remote viewing descriptions. Overall, these "blinded" judges correctly identified the target location in six of the ten trials, a result that's highly statistically significant.

Shortly after this study was completed, I conducted a replication experiment with Jo Marie Haight; this time between Durham, North Carolina, and Cape Canaveral, Florida.[5] The same design produced another significant result that supported the remote-viewing hypothesis and my own direct Akashic experience.

I'd like to say that these results from the studies with me as the viewer pleased me. Indeed, at one level they did. It was thrilling to harness these abilities within a scientific framework. But they were also deeply disturbing. I was in my early twenties and had no real grounding for these experiences. It was easier to think that other people might have these abilities than to have them myself. There was a mismatch between my abstract understanding of a paradigmatic science and my own personal experiences of the Akashic field. I have come to understand that I was not alone in my discomfort. For many people, psi experiences are unwanted and unwelcome, all the more so because we don't have a strong cultural framework in which to understand them. Gradually I began to expand my worldview to accommodate an expanded view of human possibility, including my own.

Psi in the Ganzfeld

A second striking experience came in the study I conducted with Charles Honorton using the ganzfeld paradigm, a sensory deprivation technique that stimulates visual imagery and in a certain way simulates a dream experience (many psychic experiences are reported in such states of consciousness). This testing paradigm has produced strong evidence for psi phenomena in various laboratories with many experimenters.[6] In this case, we worked with students from the Juilliard School of the Performing Arts.[7] A student was placed in an electrically shielded and soundproof room while another person was in a separate room watching a randomly selected video clip. On one occasion that I recall most vividly, I was the sender. As the experiment began, the experimenter would select a pool of four orthogonal clips from a larger set of film clips. These clips were grouped in such a way that the subject, after the session, could objectively select the clip that most strongly matched their impressions during the ganzfeld period.

The randomly selected clip that day was from the movie *Altered States*. It was the scene of a descent into hell, including a corona sun, a crucifix, and a large lizard opening and closing its mouth. As I watched in fascination, I could hear the student describing his impressions through headphones that sent his voice to me from the shielded room via a one-way communication. I can still recall the feeling of chills up and down my spine as the drama student described what I was watching, including the highly unlikely image of the lizard opening and closing his mouth at exactly the moment I was watching it on the video feed. This experience has stayed with me over the years as I have considered the nature of the evidence for psi phenomena and the debates over what is true about the limits of our consciousness. Perhaps it's this experience, more than the highly significant statistical result that we obtained, that informed my core belief in a world that includes Akashic phenomena.

The Effect of Distant Mental Intention on Living Systems

A third study that I will mention involves my work on distant intention and healing. For more than a decade, I collaborated with William Braud at the Mind Science Foundation to develop a research protocol that allows us to study the correlation between one person's intention and another person's physiology.[8] This is a procedure we eventually came to call distant mental interactions between living systems (DMILS). The idea behind the work was to simulate an experience in the laboratory that would allow us to study psychic healing, only working with healthy people who would serve as models for understanding what happens in the "real world."

Over a decade, we completed a series of process-oriented studies that resulted in a highly significant statistical deviation from chance expectation across fourteen formal, randomized, and double-blind experiments. While our results didn't prove healing per se, the work helped establish a proof of the principle that healers can affect the bodies of their patients, even at a distance.[9] In this way, we helped establish a research protocol to study what healers across the world and in many cultures believe they can do when they send healing intention to another person, even under conditions that preclude sensory exchange between them.[10]

These effects were later replicated and expanded upon in partnership with Stephen Laberge at the Cognitive Sciences Laboratory.[11] We again made use of a DMILS design, this time in regard to the remote-staring paradigm. Here we were examining the widespread experience people report of feeling someone staring at them from a distance.

The basic design involves measuring the physiology of one person while another person focuses their attention on an image sent via closed-circuit television from one room to another. The sender is instructed to send their intention during experimental periods but not during control periods. At the end of the experiment, we averaged the autonomic nervous system readings of the receiver and correlated it with the intention periods as compared to the control, or nonintention, periods. In two experiments, we again produced statistically significant evidence for a psi effect. Based on a meta-analysis, these effects have been shown to replicate in various laboratories across the world.[12]

Building on this work, in the early 1990s I began an unusual research collaboration. Working with Dr. Richard Wiseman, a psychologist, magician, and card-carrying member of the skeptical community, we began a ten-year partnership to consider the nature of the "experimenter effect."

Richard had conducted a series of DMILS studies. While my data resulted in significant deviations between the treatment and control conditions, Wiseman consistently found chance results in his studies. To help us understand why our results differed in this way, we designed two identical experiments that made use of the same laboratory, same equipment, same subject population, same randomization procedure, and so on. The only difference was that I worked with half the people and Wiseman worked with the other half.

In our first study, conducted in his laboratory at the University of Hertfordshire, we both replicated our original findings; I found a significant difference in the mean physiology of the participants between the intention and control periods, and he found a chance result.[13] This suggested that we needed to consider our assumptions about the nature of objectivity and the value of a randomized double-blind study design—both considered fundamental cornerstones of the scientific method. Perhaps consciousness needs different methods and different assumptions?

A second study was conducted in my laboratory at the Institute of Noetic Sciences to see if we could replicate our provocative findings; again we each confirmed our original results.[14]

We then designed a third study to test the hypothesis that the differences in our effects were due to a sociability factor; perhaps I was making people feel more comfortable and open than Richard was and this might explain the differences in our outcomes.[15] Unfortunately, as we conducted identical trial after trial, the project grew to be tedious and boring. Perhaps this is a factor, perhaps not. But in the end we didn't produce a significant psi effect, although there were some interesting internal effects.

I believe both of us remain curious about the differences in our experimental outcomes. We are both open about working with one another, even if our belief systems and research experiences differ.[16] Through such open-minded collaborations, we may be able to gain greater insights into the nature of Akashic experiences and how we may study them through the lens of science. It's my conviction that breakthroughs come at the points of intersection between worldviews, disciplines, and ways of knowing and being.

EXCEPTIONAL AND TRANSFORMATIVE EXPERIENCES

Over time, I have sought to understand the nature of psi and other Akashic experiences outside the laboratory. Obtaining a Ph.D. in anthropology, I felt that qualitative methods may reveal details that are left on the cutting room floor in our lab-based studies. I have been interested in how exceptional experiences impact people's lives in ways that are transformative. This has led me to engage in a decade of research on what stimulates transformation, what sustains it, and what results from experiences that open us to a larger set of possibilities.

In a recent book *Living Deeply: The Art and Science of Transformation in Everyday Life*[17] my colleagues and I use the term *transformative experience* to refer to those events that lead to lasting changes in people's lives and worldviews. We contrasted these transformative events with other extreme, extraordinary, or spiritual experiences that don't result in long-term changes in consciousness. Many people report Akashic experiences, but not all have ultimately led to deep changes in who they are and what they are capable of becoming. As transpersonal psychologist Frances Vaughn told us during an interview in 2002:

> Transformation really means a change in the way you see the world—and a shift in how you see yourself. It's not simply a change

pain & suffering

> in your point of view, but rather a whole different perception of what's possible. It's the capacity to expand your worldview so that you can appreciate different perspectives, so that you can hold multiple perspectives simultaneously. You're not just moving around from one point of view to another, you're really expanding your awareness to encompass more possibilities.[18]

We began the study by collecting people's stories about consciousness transformation. A pattern across hundreds of narratives was that true transformation is often held as a major turning point, much like the kind of hero's journey reported by Joseph Campbell.[19] We convened teachers who specialize in transformation and asked them questions that helped us to develop a language around the transformation process. We were privileged to conduct lengthy interviews with sixty masters from different transformative traditions, looking both at the differences across them as well as the patterns that connect them. Our participants represented major world religions (Christianity, Hinduism, Buddhism, Islam, Judaism), Earth-based traditions, and people who have created new forms that reflect a modern interpretation of transformation technologies. This field study led our team to craft an ecologically valid survey that has allowed us to collect data from nearly 2,000 self-selected people. The work continues today through a blend of field- and laboratory-based studies, including a current experiment to see if transformational training leads to greater interconnectedness as measured in a formal DMILS study.

Through this work, my colleagues and I have found that consciousness transformations are most commonly triggered by experiences of pain and suffering. Life events, including illness, divorce, and loss of jobs, can serve to disrupt the steady state in a person's life—giving them an opportunity, if they can see it as such, to alter their path and to live with an expanded, meaning-filled worldview. Painful and frightening experiences have the capacity to loosen our control and dissolve our identities in ways that broaden our understanding of what is possible. As physician and teacher Rachel Remen noted:

> Crisis, suffering, loss, the unexpected encounter with the unknown—all of this has the potential to initiate a shift in perspective. A way of seeing the familiar with new eyes, a way of seeing the self in a completely new way. It shuffles a person's values like a deck of cards. A value that's been on the bottom of the deck for many years turns out to be the top card. There's a moment when the individual steps away from the former life and the former identity and is completely out of control and completely surrenders—and then is reborn with a larger, expanded identity.[20]

Of course, not all catalysts of transformation are filled with pain. Many people report sensations of deep beauty, awe, wonder, and a profound connection to something greater than themselves. These Akashic experiences can often encompass what William James called mystical experiences[21] and what Abraham Maslow later referred to as peak experiences,[22] and also what Carl Jung considered to be encounters with the numinous.[23] These perceptions move us beyond our narrow definition of the self. They can take the form of a deeply rooted, embodied sense of unity, an awareness of great love, and a fundamental sense of interconnection.

In our research we have seen that transformative experiences are often sudden and profound.[24] These sudden personal metamorphoses, called "quantum changes" by Miller and C'de Baca,[25] can include psi experiences that are completely unexpected, as well as epiphanies, "big dreams," and senses of revelation. They can include various experiences that suggest an extended reach of our human consciousness, including near-death experiences, spontaneous healing, or various other abilities and phenomena that arise in nonordinary states of consciousness. Transpersonal scholar and archivist Rhea White found that, even though the phenomenology of such experiences may differ (such as seeing an apparition, sensing mystical oneness with the whole of existence, or having precognitive dreams), all these experiences can serve as a portal to a new worldview.[26]

My own worldview has been shaped by these scholars and by my experiences in and out of the laboratory. They have given me a language and a lineage line for my explorations of consciousness. For example, more than a century ago William James wrote about the transformative potential of what he defined as noetic experiences. He described them as "states of insight unplumbed by the discursive intellect."[27] These noetic forms of the Akashic experience have several essential qualities: They need to be directly experienced, for it's often the case that they aren't easily communicated to others. Second is what James called a noetic quality that makes them actual forms of knowledge. As the famed scholar noted in 1902: "They are illuminations, revelations, full of significance and importance, all inarticulate though they remain; and as a rule they carry with them a curious sense of authority."[28] They are also, according to James, states that are transient and cannot be controlled.

Now, more than a hundred years after James gave this description, my colleagues and I conduct research at the Institute of Noetic Sciences on the very phenomena he mapped out for us. We have found, for example, that more than 61 percent of the people we surveyed said their transformative experiences were due to circumstances "out of anyone's control."[29] This fact reveals yet again the complexities of bringing such experiences under the gaze of a science grounded in classical assumptions of cause and effect.

THE QUEST FOR A NEW PARADIGM

This leads me back to where I started: questing for a fundamental paradigm, cosmology, or story of the world that's inclusive enough to embrace the Akashic, noetic dimensions while not losing sight of what is real and true in the objective and intersubjective realms of lived experience.

We are alive in a time of enormous complexity—how do we make sense of the fact that a Christian, Jew, Muslim, Buddhist, secular humanist, and Pagan are all using the same grocery stores, public schools, and health care centers? How is it that a materialist scientist can sit with a contemplative monk and consider the nature of consciousness? Are there insights that come when we bridge inner and outer ways of knowing that may help us navigate twenty-first-century life and times? For me the answer is yes.

As I continue to weave together science and direct experience, I have embarked on a new project to build bridges between worldviews and ways of knowing. The program focuses on education, which is arguably the most influential societal force shaping young people today. The prevailing worldview is that the primary function of education is the development of cognitive skills. Analytical and memory skills are currently esteemed as the highest forms of intelligence (e.g., IQ). But a growing number of educators, researchers, and parents are questioning these assumptions.

Recent advances in psychology and neuroscience indicate that educating for the whole person is an idea whose time has come. From Howard Gardiner's (1983) theory of multiple intelligences[30] to the recognition of wide variability in learning styles, there is an ever-expanding view of intelligence and human potential with which our mainstream educational programs need to catch up.

What is being called for is a new model of learning that includes global students embracing a new kind of literacy that appreciates and incorporates different worldviews and ways of knowing, including Akashic knowing.

Working with a small team of researchers, educators, and scientists, we are creating a curriculum on what we are calling "worldview literacy."[31] We define this as the capacity to comprehend and communicate not only our own worldview but also to recognize that our beliefs come from our particular frame of reference and to understand that others hold different and potentially equally valid worldviews out of which their assumptions, and therefore their actions, arise. This capacity also includes being able to adapt to changes that come through a meeting of different perspectives, customs, practices, and belief systems.

The blended learning program will include multimedia presentations such as video interviews with teachers and masters of the world's cultural and religious systems, stories for children from different world traditions, video-led practices from diverse traditions, and group discussions and art. We hypothesize that we will find measurable differences before and after the program in areas like intolerance, defensiveness to difference, sense of in-group identification, ability to hold paradox, and understanding different worldviews.

The new program is grounded in the philosophy of cultural pluralism and a search for the perennial across cultures. Our goal is to help students move beyond simply tolerating diversity to developing a place of deep appreciation for our differences—as well as our points of connection. In a world that's increasingly divided, this move toward worldview literacy allows us to formulate a new paradigm that values multiple ways of knowing and being—across people and within ourselves.

FINAL REFLECTIONS

Through my work and my life experiences, I have seen that the transformative process involves a change in self that includes both our inner and our outer realities. It provides links between our direct experiences and our being in the world through action and service. Bridging the Akashic and the rational has allowed me to develop a deeper and richer sense of connection to myself, my family, my community, and my environment. In this process, I have developed an increasing awareness and appreciation for the sacred in every aspect of life.

As each of us lives into expanded human capacities, we can come back to who we are at our core level of being. Starhawk, a Wiccan teacher and writer, made this point during an interview in 2006. She noted that attending to the collective realms of interconnections, such as we find in the Akashic experience, is a way of reclaiming our natural consciousness. In her words:

> A range of different types of consciousness is available to human beings. It's a kind of anomaly that postmodern Western culture has narrowed the range of that what we are encouraged to have. Maybe it's not so much a transformation we are speaking of, but an opening. It's a reclaiming.[32]

A natural consciousness is readily available to human beings as a birthright. It is not so much a supernatural awareness as it is an awareness of being present in this world and open to understanding the interrelations and interconnections. It's about being aware and thinking in terms of patterns and relationships rather than separate isolated objects.

Finding the place within us for both Akashic insights and rational knowing is important when we seek transformation and greater self-discovery. They are also important for guiding us as we engage in the broader world. Experiences of interconnection remind us of the web that links us all, in ways both visible and invisible. Together we can build a new paradigm that embraces our wholeness, not just our parts. In this process we can cocreate a worldview that serves the fullness of who we are—individually and collectively.

SIXTEEN

Acceding to the Field

THE CASE OF NEAR-DEATH EXPERIENCES IN SURVIVORS OF CARDIAC ARREST

Pim van Lommel

Pim van Lommel, a Dutch cardiologist, has published more than twenty articles, two books, and several chapters about near-death experiences since he started his research on this subject in 1986. His book *Endless Consciousness* was published in Dutch in 2007 and subsequently in other languages. In 2005 he was awarded the Bruce Greyson Research Award on behalf of IANDS (the International Association of Near-Death Studies) in the USA, and in 2006 he received from the president of India the Life Time Achievement Award at the World Congress on Clinical and Preventive Cardiology in New Delhi.

THE EXPERIENCES

Some people who have survived a life-threatening crisis report an extraordinary conscious experience. I begin this report with the story of a woman who experienced a near-death experience (NDE) during delivery.

> Suddenly I realize I am looking down at a woman who is lying on a bed with her legs in supports. I see the nurses and doctors panicking, I see a lot of blood on the bed and on the floor, I see large

hands pressing down hard on the woman's belly, and then I see the woman giving birth to a child. The child is immediately taken to another room. I know it's dead. The nurses look dejected. Everybody is waiting. My head is knocked back hard when the pillow is pulled away. Once again, I witness a great commotion. Swift as an arrow I fly through a dark tunnel. I am engulfed by an overwhelming feeling of peace and bliss. I hear wonderful music. I see beautiful colors and gorgeous flowers in all sorts of colors in a large meadow. At the far end is a beautiful, clear, warm light. This is where I must go. I see a figure in a light garment. This figure is waiting for me and extends her hand. I feel that I am warmly and lovingly expected. We proceed hand in hand to the beautiful and warm light. Then she lets go of my hand and turns around. I feel that I am pulled back. I notice a nurse slapping me hard on my cheeks and calling me.

Once I returned from that beautiful world, that beautiful experience, my reception here in this world was cold, frosty, and above all loveless. The nurse I tried to share my beautiful experience with dismissed it by saying I would soon receive some more medication so I could sleep soundly and then it would be all over. All over? I didn't want that at all. I didn't want it to be over. I wanted to go back. The gynecologist told me I was still young, I could have plenty more children, and I should just move on and focus on the future. I stopped telling my story. Just to find words for my experience was difficult enough; how could words express what I had experienced? But what else could I do? Where could I take my story? What was the matter with me? Had I gone mad? And I kept silent. I spent years dedicated to a silent search. When, eventually, I found a book in the library with a report of an NDE, I could hardly imagine that I had had such an experience. Even I had stopped believing myself. Only very, very gradually did I come to have the courage and the strength to believe myself, to trust my experience, so I could start accepting and integrating it in my life.

As a cardiologist I have had the privilege of meeting many patients who have been willing to share their near-death experiences with me. The first time this happened was in 1969. In the coronary care unit the alarm suddenly went off. The monitor showed that the electrocardiogram (ECG) of a patient with a myocardial infarction had become flat. The man had a cardiac arrest (clinical death). After two electric shocks and a spell of unconsciousness lasting some four minutes, the patient regained consciousness, much to the relief of the nursing staff and attendant doctor. That attendant doctor was I. I had started my cardiology training that year.

Following the successful resuscitation everyone was pleased, except the patient. To everyone's surprise he was extremely disappointed. He spoke of a tunnel, of colors, of a light, of a beautiful landscape, and of music. He was extremely emotional. The term *near-death experience* didn't yet exist, nor had I ever heard of people having any recollection of the period of their cardiac arrest. In fact, while studying for my degree I had learned that such a thing is impossible: being unconscious means not being aware, and that applies to people suffering a cardiac arrest or patients in a coma. In the event of a cardiac arrest, a patient is unconscious, has stopped breathing, and has no palpable pulse or blood pressure. I had been taught that at such a moment it's simply impossible to be conscious or to have memories because all brain function has ceased.

Although I had never forgotten the successfully resuscitated patient from 1969 with his memories of the period of his cardiac arrest, I had never done anything with the experience. This changed in 1986 when I read George Ritchie's book, *Return from Tomorrow,* about near-death experiences.[1] When suffering double pneumonia as a medical student in 1943, Ritchie had experienced a period of clinical death. At the time, antibiotics such as penicillin were not yet widely used. Following a period of very high fever and extreme tightness of the chest, he passed away: he stopped breathing and his pulse had gone. He was pronounced dead by a doctor and covered with a sheet. But a male nurse was so upset by the death of this medical student that he managed to persuade the attendant doctor to administer an adrenalin injection in the chest near the heart—a most unusual procedure in those days. Having been "dead" for more than nine minutes, George Ritchie regained consciousness, to the immense surprise of the doctor and nurse.

It turned out that during his spell of unconsciousness, the period in which he had been pronounced dead, he had had an extremely deep experience of which he could recollect a great many details. At first he was quite unable and afraid to talk about it. Later he wrote his book about what happened to him in those nine minutes. And after graduation, he shared his experiences with medical students in psychiatry lectures. One of the students attending these lectures was Raymond Moody, who was so intrigued by this story that he started looking into experiences that may occur during critical medical situations. In 1975 he wrote the book *Life after Life*, which became a global bestseller.[2] In this book Moody first used the term *near-death experience*.

What is an NDE? An NDE is a special state of consciousness that occurs during an imminent or actual period of physical, psychological, or emotional death. Many circumstances are described during which NDEs are reported, such as cardiac arrest, shock after loss of blood, coma following traumatic brain injury or intracerebral hemorrhage, near-drowning (children), or asphyxia, but also in serious health conditions that are not immediately life-threatening, or depression or isolation, with or without an obvious reason. Similar experiences can occur during the terminal phase of an illness and are called deathbed visions or pre-death experiences. An NDE is the reported memory of a whole set of impressions during a nonordinary state of consciousness, including a number of special elements such as out-of-body experiences, pleasant feelings, and seeing a tunnel, a light, deceased relatives, or a life review.

NDEs are transformational, causing profound changes of life-insight and loss of the fear of death. The content of an NDE and the effects on patients seem similar worldwide, across all cultures and all times. However, the subjective nature and absence of a frame of reference for this experience lead to individual, cultural, and religious factors that determine the vocabulary used to describe and interpret the experience.

MY RESEARCH ON NDES

After reading George Ritchie's book I kept asking myself how someone could possibly experience consciousness during cardiac arrest, and indeed whether this is a common occurrence. That's why, in 1986, I started systematically asking all the patients at my out-patient clinic who had ever undergone resuscitation whether they had any recollection of the period during their cardiac arrest. I was more than a little surprised to hear, within the space of two years, twelve reports of such a near-death experience among just over fifty survivors of cardiac arrest. Prior to making these inquiries, I had not heard of such reports, except for that first time in 1969. I had not inquired after them either, because I had not been open to them. After all, according to current medical knowledge, it's impossible to experience consciousness when one's heart has stopped beating. But all these reports I started hearing roused my curiosity.

Near-death experiences are occurring with increasing frequency because of improved survival rates resulting from modern techniques of resuscitation. According to a recent randomized inquiry in Germany and the USA, about 4 percent of the total population in the Western world experiences an NDE. So more than nine million people in the USA alone must have had this experience.

Why do we physicians hardly ever hear patients tell about their near-death experiences? Patients are reluctant to share their experience with others because of the negative responses they get. As a doctor you have to be open to hear about NDEs, and patients must feel that you trust them, that you can listen without prejudice.

For me it all started with curiosity: with asking questions, with seeking to explain certain objective findings and subjective experiences. The phenomenon of near-death experience raised a number of fundamental questions. How and why does an NDE occur? How does the content of an NDE come about? Why does a person's life change so radically after an NDE? I was unable to accept some of the answers usually given to these questions, because they seemed incomplete, incorrect, or unfounded. I grew up in an academic environment in which I had been taught that there is a materialist explanation for everything. And up until the time I started really listening to my patients, I had always accepted this as indisputably true.

Some scientists don't believe in questions that cannot be answered, but they do believe in wrongly formulated questions. The year 2005 saw the publication of a special anniversary issue of the journal *Science*, featuring 125 questions that scientists had so far failed to solve.[3] The most important unanswered question, "What is the universe made of?" was followed by: "What is the biological basis of consciousness?" I would reformulate this second question as follows: "Is there a biological basis of consciousness (at all)?" We can also distinguish between both temporary and timeless aspects of our consciousness. This prompts the following question: "Is it possible to speak of a beginning of our consciousness, and will our consciousness ever end?"

In order to answer these questions, we need a better understanding of the relationship between brain function and consciousness. We shall have to start by examining whether there is any indication that consciousness can be experienced during sleep, coma, brain death, clinical death, the process of dying, and, finally, after confirmed death. If the answers to any of these questions are positive, we must look for scientific explanations and scrutinize the relationship between brain function and consciousness in these different situations. By studying everything that has been thought and written about death throughout history, in all times, cultures, and religions, we may be able to form a different or better picture of death. But we may achieve the same on the basis of findings from recent scientific research into near-death experiences.

It has emerged that most people lose all fear of death after an NDE. Their experience tells them that death is not the end of everything and that life goes on in one way or another. According to people with an NDE, death is nothing other than a different way of being with an enhanced and broadened consciousness, which is everywhere at once because it's no longer tied to a body. This is what someone wrote to me after his NDE:

It's outside my domain to discuss something that can only be proved by death. However, for me personally this experience was decisive in convincing me that consciousness endures beyond the grave. Death turned out to be not death, but another form of life.

THE DUTCH PROSPECTIVE STUDY

In order to obtain more reliable data to corroborate or refute the existing theories on the cause and content of an NDE, we needed a properly designed scientific study. This was the reason why in 1988 Ruud van Wees and Vincent Meijers, both psychologists who wrote their doctoral theses on NDE, and I, a cardiologist with an interest in the subject, started designing a prospective study in the Netherlands. At that point, no large-scale prospective studies into NDE had been undertaken anywhere in the world. Our study aimed to include all consecutive patients in the ten participating hospitals who survived a cardiac arrest. In other words, all the patients in our study had been clinically dead for a short period of time. Clinical death is defined as the period of unconsciousness caused by lack of oxygen in the brain (anoxia) because of the arrest of circulation, breathing, or both, as caused by cardiac arrest in patients with an acute myocardial infarction. In this situation, if no resuscitation is initiated, the brain cells will be irreversibly damaged within five to ten minutes and the patient will die.

In a prospective study all resuscitated patients are asked, within a few days of their resuscitation, whether they have any recollection of the period of their cardiac arrest, that is, of the period of their unconsciousness. All patients' medical and other data are carefully recorded before, during, and after their resuscitation. We had a record of the electrocardiogram, or ECG, for all patients included in our study. An ECG displays the electrical activity of the heart. In cardiac arrest patients this ECG record always displays a usually lethal arrhythmia (ventricular fibrillation) or an asystole (a flat line on the ECG). In the event of resuscitation outside the hospital we were given the ECG done by the ambulance staff. Following successful resuscitation we carefully recorded the demographic data of all patients, including age, sex, education, religion, foreknowledge of NDE, and whether they had had an earlier NDE. They were also asked whether they had been afraid of dying just before their cardiac arrest. Likewise, we carefully recorded all medical information, such as the duration of the actual cardiac arrest and the duration of unconsciousness, how often the patient required resuscitation, and what medication, in what dosage, was administered before, during, and after resuscitation. We also recorded how many days after resuscitation the interview took place, whether the patient was lucid during the interview, and whether his or her short-term memory was functioning well.

Within four years, between 1988 and 1992, 344 successive patients who had undergone a total of 509 successful resuscitations were included in the study. Our study design created a control group of patients: those who survived a cardiac arrest but who had no recollection of the period of unconsciousness. A longitudinal study into life changes was based on interviews after two and eight years with all patients who had reported an NDE and who were still alive, as well as with a control group of post-resuscitation patients who were matched for age and sex, but who had not reported an NDE. Our primary question was whether the customary changes in attitude to life after an NDE were the result of surviving a cardiac arrest or whether these changes were caused by the experience of an NDE. This question had never been subject to scientific and systematic research before. The Dutch study was published in *The Lancet* in December 2001.[4]

When patients reported memories from the period of unconsciousness, their experiences were scored according to the WCEI: "weighted core experience index."[5] The higher the number of elements reported, the higher the score and the deeper the NDE. Our study found that of the 344 patients, 282 (82 percent) had no recollection of the period of their unconsciousness, whereas 62 patients—18 percent of the 344 patients—reported an NDE. Of these 62 patients with memories, 21 patients had some recollection; having experienced only some elements, they had a superficial NDE with a low score. Eighteen patients had a moderately deep NDE, seventeen patients reported a deep NDE, and six patients a very deep NDE.

Half of the patients with an NDE were aware of being dead and had positive emotions; 30 percent had a tunnel experience, observed a celestial landscape, or met with deceased persons; approximately 25 percent had an out-of-body experience, communication with "the light," or perception of colors; 13 percent had a life review; and 8 percent experienced the presence of a boundary. In other words, all the familiar elements of an NDE were reported in our study, with the exception of a frightening or negative NDE.

Are there any reasons why some people do but most people don't recollect the period of their unconsciousness? In order to answer this question we compared the recorded data of the 62 patients with an NDE with the data of the 282 patients without an NDE. To our surprise we didn't find any significant differences in the duration of the cardiac arrest, no differences in the duration of unconsciousness, and no differences in whether intubation was necessary for artificial respiration in seriously ill patients who remained in a coma for days or weeks after a complicated resuscitation. Nor did we find differences in the thirty patients who had a cardiac arrest during electrophysiological stimulation (EPS) in the catheterization laboratory and whose heart rhythms were reestablished via defibrillation (an electric shock) within fifteen to thirty seconds. Thus we failed to identify differences between patients based on the length of the period of cardiac arrest. The degree or gravity of the lack of oxygen in the brain (anoxia) appeared to be irrelevant. Likewise, it was established that medication played no role. Most patients suffering a myocardial infarction receive morphine-type painkillers, while people who are put on a respirator following complicated resuscitation are given extremely high doses of sedatives.

Psychological causes such as the infrequently noted fear of death did not affect the occurrence of an NDE either, although it did affect the depth of the experience. Whether or not patients had heard or read anything about NDE in the past didn't make a difference either. Any kind of religious belief, or indeed its absence in nonreligious people or atheists, was irrelevant, and the same was true for the level of education. Factors that *did* affect the frequency of an NDE were being below sixty years of age and having several resuscitations during the stay in the hospital. In these cases the chances of a report of an NDE were greater. Remarkably, we found in our study that patients who had had an NDE in the past reported NDEs significantly more frequently. A complicated resuscitation can result in a long coma, and most patients who have been unconscious on a respirator for days or weeks are more likely to suffer short-term memory defects as a result of permanent brain damage. These patients reported significantly fewer NDEs in our study. This suggests that a good memory is essential for remembering an NDE.

We were particularly surprised to find no medical explanation for the occurrence of NDEs. All the patients in our study had been clinically dead, and only a small percentage reported an enhanced consciousness with lucid thoughts, emotions, memories, and sometimes perception from a position outside and above the lifeless body while doctors and nursing staff were carrying out resuscitation. If there were a physiological explanation for the occurrence of this enhanced consciousness, such as a lack of oxygen in the brain (anoxia), one might have expected all patients in our study to report an NDE. They had all been unconscious as a result of their cardiac arrest, which caused the loss of blood pressure and the cessation of breathing and all physical and brain-stem reflexes. Likewise the gravity of the medical situation, such as long-term coma after a complicated resuscitation, failed to explain why patients did or didn't report an NDE, except in the case of lingering memory defects.

The psychological explanation is doubtful because most patients didn't experience any fear of death during their cardiac arrest, it having occurred so suddenly that they failed to notice it. In most cases they were left without any recollection of their resuscitation. This is borne out by Greyson's study,[6] which only collected the subjective data of patients after their resuscitation and showed that most patients didn't even realize they had had a cardiac arrest. This is similar to fainting. When people regain consciousness they have no clear idea of what happened. A pharmacological explanation could be excluded as well, as the medications administered had no effect on whether patients reported an NDE.

Elements of NDEs that were mentioned by patients in our study include an out-of-body experience, a holographic life review, a meeting with deceased relatives, and a conscious return into the body. These elements were experienced during the period of cardiac arrest, the period of apparent unconsciousness: the period of clinical death.

THE OUT-OF-BODY EXPERIENCE

In this experience people have veridical perceptions from a position outside and above their lifeless body. The out-of-body experience (OBE) is scientifically important because doctors, nurses, and relatives can verify the reported perceptions, and they can also corroborate the precise moment when the NDE with OBE occurred during the period of cardiopulmonary resuscitation (CPR). This proves that OBE cannot be a hallucination, because a hallucination is the experience of a perception that has no basis in "reality." It cannot be a delusion, which is an incorrect assessment of a correct perception, nor an illusion, which means a misapprehensive or misleading image.

Should an OBE be considered a kind of nonsensory perception—a variety of Akashic experience? This is the report of a nurse at a coronary care unit:

> During the night shift an ambulance brought a forty-four-year-old cyanotic, comatose man into the coronary care unit. He had been found in coma about thirty minutes before in a meadow. When we tried to intubate the patient, we found that he had dentures in his mouth. I removed his upper dentures and put them onto the "crash cart." After about an hour and a half the patient had sufficient heart rhythm and blood pressure, but he was still ventilated and intubated, and he was still comatose. He was transferred to the intensive care unit to continue the necessary artificial respiration. I met

the patient again only after more than a week. By then he was in the cardiac ward. The moment he saw me he said: "Oh, that nurse knows where my dentures are." I was very, very surprised. Then he elucidated: "You were there when I was brought into the hospital, and you took my dentures out of my mouth and put them into that cart; it had many bottles on it and a sliding drawer underneath, and there you put my teeth." I was especially amazed because I remembered this happening while the man was in deep coma and we were in the process of administering CPR. It appeared that the man had seen himself lying in bed, and that he had perceived from above how we nurses and doctors had been busy with the CPR. He was also able to describe correctly and in detail the small room in which he had been resuscitated as well as the appearance of those present, like myself.

THE HOLOGRAPHIC LIFE REVIEW

Here the subject feels the presence of, and re-experiences, not only every act but also every thought from his or her life and realizes that all is an energy field that influences others as well. All that has been done and thought seems to be stored. Because of their connection with the memories, emotions, and consciousness of other persons, subjects experience the consequences of their own thoughts, words, and actions for those persons at the very moment when they occurred. They understand that everything one does to others will ultimately be returned to oneself. Patients survey their whole life in one glance; time and space don't seem to exist during this experience—instantaneously they are where they focus their attention. They can talk for hours about the content of their life review even though resuscitation only took minutes.

Here are some statements:

> All of my life up till the present seemed to be placed before me in a kind of panoramic, three-dimensional review, and each event seemed to be accompanied by a consciousness of good or evil or with an insight into cause or effect.

Not only did I perceive everything from my own viewpoint, but I also knew the thoughts of everyone involved in the event, as if I had their thoughts within me. This meant that I perceived not only what I had done or thought, but even in what way it had influenced others, as if I saw things with all-seeing eyes. And so even your thoughts are apparently not wiped out. And all the time during the review the importance of love was emphasized.

Looking back, I cannot say how long this life review and life insight lasted, it may have been long, for every subject came up, but at the same time it seemed just a fraction of a second, because I perceived it all at the same moment. Time and distance seemed not to exist. I was in all places at the same time, and sometimes my attention was drawn to something, and then I would be present there.

A *preview* (flash forward) was also sometimes experienced, in which future images from personal life events as well as more general images of the future occur. Again, it seems as if time and space don't exist during this experience. Some patients describe how they returned into their bodies, mostly through the top of the head, after they had come to understand that "it wasn't their time yet" or that "they still had a task to fulfill." This conscious return into the body is experienced as something very oppressive. Patients realize that they are "locked up" in their damaged bodies, and they re-encounter all the pain and limitations of their conditions.

The subsequent interviews in the longitudinal study were conducted using a standardized inventory featuring thirty-four life-change questions.[7] Among the seventy-four patients who consented to be interviewed after two years, thirteen of the total thirty-four factors listed in the questionnaire turned out to be significantly different for those with and those without an NDE. The interviews showed that in people with an NDE the fear of death had significantly decreased, while belief in an afterlife had significantly increased.

After eight years we again compared the thirteen factors that after two years had been significantly different between the two groups with and without NDEs. It struck us that both groups were undergoing unmistakable processes of transformation. We were surprised to find that the processes of transformation that had begun after two years in persons with an NDE had clearly intensified after eight years. And the same was true for those without an NDE. While clear differences remained between people with and without NDEs, these differences had become less marked.

In summary, we could say that eight years after their cardiac arrest, all patients had changed in many respects, showing more interest in nature, the environment, and social justice, displaying more love and emotions and being more supportive and involved in family life.

Nevertheless the people who had experienced an NDE during their cardiac arrest remained clearly different. They continued to be less afraid of death and had stronger beliefs in an afterlife. We saw that they had a greater interest in spirituality and the purpose of life, as well as a greater acceptance of and love for themselves and others. Likewise, they displayed a greater appreciation of ordinary things, whereas their interest in possessions and power had decreased.

The conversations also revealed that people had acquired enhanced intuitive feelings after an NDE, along with a strong sense of connectedness with others and with nature. As many of them put it, they had acquired "paranormal gifts." Many observed that the sudden occurrence of enhanced intuition had been quite problematic, as they suddenly had a very acute sense of others, which can be extremely intimidating; they also experienced clairvoyance, prophetic feelings, and visions. This intuitive sense can be quite extreme, with people "sensing" feelings and sadness in others, or having the sense of knowing when someone will die—which usually proves accurate. The integration and acceptance of an NDE is a process that can take many years because of its far-reaching impact on the person's pre-NDE understanding of life and values. It's quite remarkable to see a cardiac arrest that lasts just a few minutes give rise to a lifelong process of transformation.

An NDE is both an existential crisis and an intense lesson in life. An NDE gives a person a conscious experience of a dimension in which time and distance play no role, in which past and future can be glimpsed, where they feel complete and healed, and where they experience unlimited knowledge and unconditional love. The life changes that follow mainly spring from the insight that love and compassion for oneself, for others, and for nature are major prerequisites of life. Following an NDE people realize that everything and everyone is connected, that every thought has an effect on both oneself and others, and that our consciousness continues beyond physical death.

PUZZLES AND POSSIBLE EXPLANATIONS

The large-scale Dutch study permitted a statistical analysis of the factors related to the occurrence of an NDE. We were able to rule out physiological, psychological, and pharmacological explanations. Our study was also the first to include a longitudinal component with interviews after two and eight years, which allowed us to compare the processes of transformation in people with and without an NDE. We identified a distinct pattern of change in people with an NDE and revealed that integrating these changes into daily life is a long and arduous process. And we reached the inevitable conclusion that patients experienced all these NDE elements during the period of their cardiac arrest, while there was a total cessation of blood supply to their brain. We were unable to answer the question regarding how this could be possible.

Bruce Greyson, who published a prospective study of 116 survivors of cardiac arrest in the USA, wrote that

> no one physiological or psychological model by itself could explain all the common features of an NDE. The paradoxical occurrence of a heightened, lucid awareness and logical thought processes during a period of impaired cerebral perfusion raises particular perplexing questions for our current understanding of consciousness and its relation to brain function. A clear sensorium and complex perceptual processes during a period of apparent clinical death challenge the concept that consciousness is localized exclusively in the brain.[8]

The British prospective study by Sam Parnia and Peter Fenwick included 63 patients who survived cardiac arrest. According to them "the NDE-reports suggest that the NDE occurs during the period of unconsciousness." This is a surprising conclusion, because

> when the brain is so dysfunctional that the patient is deeply comatose, those cerebral structures, which underpin subjective experience and memory, must be severely impaired. Complex experiences such as are reported in the NDE should not arise or be retained in memory. Such patients would be expected to have no subjective experience, as was the case in the vast majority of patients who survive cardiac arrest, since all centers in the brain that are responsible for generating conscious experiences have stopped functioning as a result of the lack of oxygen.[9]

A frequently cited explanation is that the observed experiences occur during the early phases of the cessation of consciousness or during its recovery. Parnia and Fenwick claim, however, that the verifiable elements of an out-of-body experience during unconsciousness, such as patients' reports on their resuscitation, render this extremely unlikely. Over a period of four years Penny Sartori carried out a smaller study of NDE in thirty-nine survivors of cardiac arrest in the UK. She concluded that "according to mainstream science, it's quite impossible to find a scientific explanation for the NDE as long as we 'believe' that consciousness is only a side effect of a functioning brain."[10] The fact that people report lucid experiences in their consciousness when brain activity has ceased is, in her view, difficult to reconcile with current medical opinion.

With current medical and scientific concepts it seems impossible to explain all aspects of the subjective experiences as reported by patients with an NDE during a transient loss of all brain functions. Furthermore, even blind people have described veridical perceptions during out-of-body experiences at the time of their NDE. It's indeed a challenge to science to come up with new hypotheses that could explain the occurrence of the following during a period of impaired cerebral perfusion, resulting in a flat EEG within fifteen seconds on the average:

- the reported interconnectedness with the consciousness of other persons and of deceased relatives
- the possibility of experiencing instantaneously and simultaneously a review and a preview of one's life in a dimension without our conventional body-linked concept of time and space, where all past, present, and future events exist
- the possibility of having clear consciousness with memories, with self-identity, with cognition, and the possibility of perception[11]

The quite often proposed objection that a flat line EEG does not rule out all brain activity, because it's mainly a registration of electrical activity of the cerebral cortex, misses the mark. The issue is not whether there is any brain activity of any kind whatsoever, but whether there is brain activity of the specific form regarded by contemporary neuroscience as the necessary condition of conscious experience. And it has been proved that there is no such specific brain activity during cardiac arrest. Moreover, despite measurable EEG activity in the brain recorded during deep sleep (no-REM phase), no consciousness is experienced because there is no integration of information and no communication between the different neural networks.[12] So even in circumstances where brain activity can be measured, sometimes no consciousness can be experienced.

In some articles[13] and in my recent book[14] I describe a concept in which our whole and undivided consciousness with declarative memories finds its origin in, and is stored in, a nonlocal space as wave-fields of information. This nonlocal aspect of consciousness can be compared to gravitational fields, where only the physical effects can be measured, and the fields themselves are not directly demonstrable. In this concept the cortex only serves as a relay station for parts of these wave-fields of consciousness to be received into or as our waking consciousness. The latter belongs to our physical body.

Thus there are two complementary aspects of consciousness, which cannot be reduced to one another, and the neuronal networks should be regarded as functioning as receivers and conveyors, not as retainers of consciousness and memories. In this concept, consciousness is not rooted in the measurable domain of physics, our manifest world. The wave aspect of our indestructible consciousness in nonlocal space is inherently incapable of being measured by physical means. However, the physical aspect of consciousness, which originates from the wave aspect of our consciousness, can be measured by means of neuro-imaging techniques like EEG, fMRI, and PET-scan.

Based on my NDE research, I conclude that our waking consciousness, which we experience as our daily consciousness, is only a complementary aspect of our whole and infinite nonlocal consciousness. This consciousness is based on indestructible and constantly evolving fields of information, where all knowledge, wisdom, and unconditional love are present and available, and these fields of consciousness are stored in a dimension beyond our concept of time and space with nonlocal and universal interconnectedness. One could call this our higher consciousness, divine consciousness, or cosmic consciousness. It's the Akashic field to which conditions at the portals of death provide a special kind of access.

SEVENTEEN

Evidence for the Akashic Field from Modern Consciousness Research

Stanislav Grof

Stanislav Grof is a psychiatrist with more than fifty years of experience in research of nonordinary states of consciousness induced by psychedelic substances and various nonpharmacological methods. He is Professor of Psychology at the California Institute of Integral Studies (CIIS) in San Francisco, conducts professional training programs, and gives lectures and seminars worldwide. He has published numerous books and more than 140 papers in professional journals. Grof is one of the founders of transpersonal psychology and the founding president of the International Transpersonal Association (ITA). In October 2007 he received the Vision 97 Award from the Dagmar and Vaclav Havel Foundation in Prague.

In the course of the twentieth century, various disciplines of modern science have amassed an extraordinary array of observations, which cannot be accounted for or adequately explained in terms of the materialistic worldview. These "anomalous phenomena" came from a wide range of fields from astrophysics, quantum-relativistic physics, and chemistry to biology, anthropology, thanatology, parapsychology, and psychology.

Pioneering scientists from various scientific disciplines have made more or less successful attempts to tackle the formidable conceptual problems presented by the anomalous data. They have formulated theories, which have suggested revolutionary new ways of looking at the recalcitrant problems they faced in their respective fields. Over time, a radically different understanding of reality and of human nature started to come into view, which is usually referred to as the new or emerging paradigm in science.

However, this new perspective constituted a mosaic consisting of impressive but disconnected pieces—Barrow and Tipler's anthropic principle, philosophical implications of quantum-relativistic physics, David Bohm's theory of holomovement, Karl Pribram's holographic model of the brain, Rupert Sheldrake's theory of morphogenetic fields, Ilya Prigogine's theory of dissipative structures, and others.[1] The credit for finding an elegant interdisciplinary solution for the anomalies plaguing modern science and for bringing these disjointed efforts together belongs to Ervin Laszlo, arguably the world's greatest system theorist and interdisciplinary philosopher of science.

Laszlo achieved that by formulating his "connectivity hypothesis," the main cornerstone of which is the existence of what he earlier called the psi field and more recently the Akashic field, a subquantum field in which everything that has ever happened remains permanently holographically recorded.[2]

NOTES FROM MY RESEARCH ON AKASHIC EXPERIENCES

My primary area of interest and my lifetime passion has been the study of an important subcategory of nonordinary states of consciousness, which I call *holotropic* (literally, "moving toward wholeness," from the Greek *holos* = "whole" and *trepein* = "moving toward"). What makes these states a fascinating object of study is their heuristic, therapeutic, and evolutionary potential.[3] This research I have been involved in for more than half a century has generated paradigm-breaking anomalous phenomena.

Holotropic states cover a rich spectrum from those that shamans experience in their initiatory crises and use in their healing practices, native people induce in rites of passage and healing ceremonies, and initiates experienced in ancient mysteries of death and rebirth, to extraordinary experiences that occur in the systematic spiritual practices of yogis, Buddhists, Taoists, Kabbalists, Sufis, Christian hesychasts, and Desert Fathers, as well as those that are found in the accounts of mystics of all countries and historical periods. Modern psychiatrists and therapists have encountered these states in psychedelic therapy, in deep experiential work without the use of psychoactive substances, in experiments with sensory deprivation and biofeedback, and during therapy of individuals undergoing spontaneous episodes of holotropic states.

The five decades that I have dedicated to consciousness research have been for me an extraordinary adventure of discovery and self-discovery. I spent approximately half of this time conducting therapy with psychedelic substances, first in Czechoslovakia, in the Psychiatric Research Institute in Prague, and then in the United States, at the Maryland Psychiatric Research Center in Baltimore, where I participated in the last surviving official American psychedelic research program. Since 1975, I have worked with holotropic breathwork, a powerful method of therapy and self-exploration that I have developed jointly with my wife, Christina. Over the years, we have also supported many people undergoing spontaneous psychospiritual crises, or "spiritual emergencies" as Christina and I call them.[4]

In my work with holotropic states, I encountered practically daily phenomena that could not be explained by the theoretical frameworks of academic psychiatry and that represented a serious challenge to the basic metaphysical assumptions of Western science. My early observations in psychedelic therapy showed beyond any reasonable doubt that my clients were able to relive in holotropic states their own biological birth and even episodes from prenatal life, often with extraordinary details; many of these experiences could be objectively verified. In academic psychiatry, the memory of birth is considered impossible, because the cerebral cortex of the newborn is not fully myelinized. I have shown the weakness of this argumentation in other publications.[5]

But I also encountered much more formidable conceptual challenges. In many instances, my clients experienced episodes from the lives of their ancestors and even animal ancestors. The only conceivable material substrate for experiences from times that precede our conception would have to be contained in the nucleus of a single cell—the sperm or the ovum—or, more specifically, in the DNA. However, we have observed experiences for which even this rather fantastic possibility would not provide a satisfactory explanatory principle. We have witnessed many situations where the experiences crossed biological hereditary lines. People often experientially identified with persons from a different racial group, such as a Slavic person with a Japanese samurai, an Anglo Saxon with a black African slave, or a Japanese with a Spanish conquistador, and so on.

Similarly, experiential identification with animals was not limited to our direct animal ancestors, where the germinal cells could—at least theoretically—serve as the material carrier of the information. We have, for example, seen in our experimental subjects authentic experiential identification with a silverback gorilla or a chimpanzee, animals with which we have common ancient protohominid ancestors but not a connection through a direct genetic line. The Darwinian evolutionary tree branches into the gorilla and chimpanzee lines long before it reaches *Homo sapiens*. This problem is even more obvious if such identification involves birds, reptiles, or insects.

It is important to emphasize that all the above types of experiential identification with other human beings, animals, and plants typically provide detailed and accurate information about various aspects of the world that reaches far beyond what the individuals could have obtained through conventional channels in the course of their present lifetime. The validation of the information made available through experiential identification is easier when it involves animals. The reason for it is that there are many important aspects of these experiences that cannot be conveyed by conventional media—books, photographs, movies, television, and so on. It is common, for example, to experience a full body image of the species involved, its state of consciousness, specific perception of the world reflecting the anatomy of its optical system (e.g., that of a raptor bird, bee, or octopus) or its acoustic system (dolphin or bat), and so on. These experiences also often provide extraordinary insights concerning animal psychology, ethology, courtship dances, or mating habits.

Ancestral, racial, and collective memories and past incarnation experiences frequently provide very specific details about architecture, costumes, weapons, art forms, social structure, and religious or ritual practices of the culture and historical period involved and sometimes even specific historical events. The veracity of this information is not always easy to assess, as it requires very accurate description of the events involved, as well as access to relevant family chronicles, archives, or historical sources. Yet there are exceptional cases where these criteria are met and the results of the verification process can be truly astonishing.

In a previous publication, I explored the groundbreaking contributions that Ervin Laszlo's work has made to two fields in which I have professional experience—consciousness research and transpersonal psychology.[6] The scope of that paper on Ervin Laszlo's work did not allow me to include practical examples of the observations that I was writing about. The present publication makes it possible for me to include case histories, which make this material come alive.

STORIES SURFACING IN MY RESEARCH

I give here concrete examples of experiences that reached far back into history and contained specific details that could be independently verified. The first story takes us to seventeenth-century Europe, to the beginning of the Thirty Years' War. This case illustrates extremely well the conceptual challenges associated with the attempts to verify the historical material from holotropic states and to identify the channels through which the information is transmitted.

Ancestral Memory or Past Life Experience? The Story of Renata

The protagonist in this story is Renata, a former client of mine in Czechoslovakia, who came to treatment because she was suffering from cancerophobia, which was making her life difficult. In a series of LSD sessions, she relived various traumatic experiences from her childhood and repeatedly dealt with the memory of her birth. In the advanced stage of her self-exploration, the nature of her sessions suddenly dramatically changed. What happened was very unusual and unprecedented.

Four of her consecutive LSD sessions almost exclusively focused on material from a specific historical period. She experienced a number of episodes that took place in seventeenth-century Prague, at a crucial period in Czech history. After the disastrous battle of White Mountain in 1621, which marked the beginning of the Thirty Years' War in Europe, the country ceased to exist as an independent kingdom and came under the hegemony of the Hapsburg dynasty. In an effort to destroy the feelings of national pride and to defeat the forces of resistance, the Hapsburgs sent out mercenaries to capture the country's most powerful noblemen. Twenty-seven prominent aristocrats were arrested and beheaded in a public execution on a scaffold erected in the Old Town Square in Prague.

During her historical sessions, Renata had an unusual variety of images and insights concerning the architecture of the experienced period and typical garments and costumes, as well as weapons and various utensils used in everyday life. She was also able to describe many of the complicated relationships existing at that time between the royal family and the vassals. Renata had never specifically studied this period of Czech history nor had she been interested in it. I had to go to the library and do historical research in order to confirm that the information Renata reported was accurate.

Many of Renata's experiences were related to various periods in the life of a young nobleman, one of the twenty-seven aristocrats beheaded by the Hapsburgs. In a dramatic sequence, she finally relived with powerful emotions and in considerable detail the actual events of the execution, including this nobleman's terminal anguish and agony. On many occasions, Renata experienced full identification with this individual. She was not able to figure out how these historical sequences were related to her present life, why they emerged in her therapy, or what they meant. After much reflection, Renata finally concluded that she must have relived events from the life of one of her ancestors. All this happened at an early stage of my psychedelic explorations, and I, admittedly, was not quite intellectually ready for this interpretation.

Trying to reach some understanding, I chose two different approaches. On the one hand, I spent a considerable amount of time in an effort to verify the specific historical information involved and was increasingly impressed by its accuracy. On the other hand, I tried to use the Freudian method of free associations, treating Renata's story as if it were a dream. I hoped that I would be able to decipher it as a symbolic disguise for some childhood experiences or problems in her present life. No matter how hard I tried, the experiential sequences did not make much sense from a psychoanalytic point of view. When Renata's LSD experiences moved into new areas, I finally gave up, stopped thinking about this peculiar incident, and focused on other more recent and immediate conceptual challenges.

Two years later, after I had come to the United States, I received a long letter from Renata with the following unusual introduction: "Dear Dr. Grof, you will probably think that I am absolutely insane when I share with you the results of my recent private search." In the text that followed, Renata described how she happened to meet her father, whom she had not seen since her parents' divorce when she was three years old. After a short discussion, her father invited her to have dinner with him, his second wife, and their children. After dinner, he told her that he wanted to share with her something she might find interesting.

In WW II, the Nazis issued an order requesting that all families in the occupied territories present to the German authorities their pedigrees demonstrating the absence of any persons of Jewish origin for the last five generations. This was a very serious issue, since failure to prove the "purity" of the family lineage had catastrophic consequences for its members. While conducting this mandatory genealogical research, Renata's father became fascinated by this procedure. After he had completed the required five-generation pedigree for the authorities, he continued this quest because of his private interest.

He was able to trace back the history of his family more than three centuries, thanks to the meticulously kept archives of the European parish houses that had preserved birth records of all the people born in their district for untold generations. He was able to show Renata the fruit of many years of his investigation, a carefully designed, complex pedigree of their family, indicating that they were descendants of one of the noblemen executed after the battle of White Mountain in the Old Town Square in Prague.

Renata was astonished by this unexpected confirmation of the information she had obtained in her LSD sessions. After having described this extraordinary episode, she expressed her firm belief that "highly emotionally charged memories could be imprinted in the genetic code and transmitted through centuries to future generations." Renata's letter ended with a triumphant "I told you so." She felt that this new unexpected information provided by her father confirmed what she had suspected all along on the basis of the convincing nature of her experiences—that she had encountered an authentic ancestral memory. As I mentioned earlier, this was a conclusion I was at the time reluctant to accept.

After my initial astonishment concerning this most unusual coincidence, I discovered a rather serious logical inconsistency in Renata's account. One of the experiences she had had in her historical LSD sessions involved the execution of the young nobleman, including all the sensation and emotions associated with the beheading. In the seventeenth century, long before the revolutionary breakthroughs of modern obstetrics, a dead person was not able to procreate. Death would have destroyed all material channels through which any information about the life of the deceased could be transmitted to posterity.

As a result of this realization, the situation got even more complicated than it was before: "the plot got thicker." On the one hand,

Renata's experience received a powerful independent validation from her father's genealogical research. On the other, there was no material substrate to account for the storage, transmission, and retrieval of the information involved. However, before we discard the information contained in Renata's story as supportive evidence for the authenticity of ancestral memories, several facts deserve serious consideration.

None of the other Czech patients, who had a total of more than two thousand sessions, had ever even mentioned this historical period. In Renata's case, four consecutive LSD sessions contained, almost exclusively, historical sequences from this time. And the probability that the convergence of Renata's inner quest and her father's genealogical research was a meaningless coincidence is so astronomical that it's difficult to take this alternative seriously. Clearly, academic psychiatry has no sober logical explanation for this concatenation of events.

Occasionally, people who have read or heard this story have suggested that Renata might have picked up the information from her father, who was passionately involved in his genealogical quest. However, Renata's knowledge of the nobleman's life and of the historical period went far beyond her father's knowledge of the same. While he was excited about his genealogical finding, his enthusiasm had not inspired a genuine historical interest and quest. We are thus left with an extraordinary situation for which the current materialistic paradigm has no explanation. It is an example of the observations from modern consciousness research that, through the work inter alia of Stanley Krippner, have recently received the name "anomalous phenomena."

The Siege of Dún an Óir: The Story of Karl

Past life memories represent a fascinating and clinically important group of experiences in holotropic states. Besides bringing new accurate information about other cultures and other historical periods, they often provide important insights into various otherwise incomprehensible areas of the person's present life, and have a remarkable healing potential. As impressive and convincing as the various features of past life experiences may be, in and of themselves, the dream of every researcher in this area is to find cases where some important aspects of these experiences can be verified by independent historical research. For me such dream came true when Christina and I met Karl and had the privilege to facilitate his process of deep self-exploration and healing. Karl enrolled in one of our Esalen month-long seminars after he had done some inner work in a renegade primal therapy group in Canada. It was one of the groups of people who had left the institute of primal therapy in Los Angeles after serious disagreements with Arthur Janov.

In the course of primal therapy, these people started having various forms of transpersonal experiences, such as archetypal visions, identification with various animals, and past life memories. Janov, who had no understanding of the transpersonal domain of the unconscious, was violently opposed to anything related to spirituality and interpreted these experiences as "copping-out from primal pain." Many people, who valued the technique of primal therapy but could not stand the straitjacket of Janov's conceptual prejudice, left his institute and formed their own groups.

Karl had begun his self-exploration as a member of such a group. After some time, his inner process reached the perinatal level. As he was reliving various aspects of his biological birth, he started experiencing fragments of dramatic scenes that seemed to be happening in another century and in a foreign country. They involved powerful emotions and physical feelings and seemed to have deep and intimate connection with his life; yet none of them made any sense in terms of his present biography. He had visions of tunnels, underground storage spaces, military barracks, thick walls, and ramparts, which all seemed to be parts of a fortress situated on a rock overlooking an ocean shore. This was interspersed with images of soldiers in a variety of situations. He felt puzzled, since the soldiers seemed to be Spanish, but the scenery looked more like Scotland or Ireland.

This was the time when Karl came to our Esalen workshop and shifted from primal therapy to holotropic breathwork. As the process continued, the scenes were becoming more dramatic and involved; many of them represented fierce combat and bloody slaughter. Although surrounded by soldiers, Karl experienced himself as a priest and at one point had a very moving vision that involved a Bible and a cross. At this point, he saw a seal ring on his hand and could clearly recognize the initials that it bore.

Being a talented artist, he decided to document this strange process, although he did not understand it at the time. He produced a series of drawings and very powerful and impulsive finger paintings. Some of these depicted different parts of the fortress, others scenes of slaughter, and a few of them Karl's own experiences, including being gored by a sword of a British soldier, thrown over the ramparts of the fortress, and dying on the shore. Among these pictures was a drawing of his hand with the seal ring engraved with the initials of the priest's name.

As he was recovering bits and pieces of this story, Karl was finding more and more meaningful connections between various aspects of its plot and his present life. He started suspecting that the drama of the Spanish priest in remote past might be the source of many of his own emotional and psychosomatic symptoms, as well as interpersonal problems. A turning point came when Karl suddenly decided on an impulse to spend his holiday in Ireland. After his return, when he was looking at the slides he had shot on Ireland's western coast, he realized that he had taken eleven consecutive pictures of exactly the same scenery. This surprised him, since he did not remember having done it and the view he had chosen was not particularly interesting.

Being a pragmatic man, he took a very rational and analytic approach to this quizzical situation. He looked at the map and reconstructed where he stood at the time and in which direction he was shooting. He discovered that the place that had attracted his attention was the ruin of an old fortress called Dún an Óir, or Forte de Oro (Golden Fortress). From the distance he was shooting, it was barely visible with the naked eye and he had to look hard to find it in the slide. Suspecting a connection between his strange behavior and his experiences from primal therapy and holotropic breathwork, Karl decided to study the history of Dún an Óir, looking for any possible clues.

He discovered, to his enormous surprise, that in 1580, a small invasion force of Spanish soldiers had landed in the nearby Smerwick Harbor to assist the Irish in the Desmond Rebellion. After being joined by some Irish soldiers they numbered about six hundred. They managed to garrison themselves within the defenses of the fort at Dún an Óir, before they were surrounded and besieged by a larger English force commanded by Lord Grey. Walter Raleigh, who accompanied Lord Grey, played the role of mediator in this conflict and negotiated with the Spaniards. He promised them free egress from the fortress, if they would open the gate and surrender to the British. The Spaniards agreed to accept this condition and surrendered, but the British didn't hold their promise. Once inside the fortress, they mercilessly slaughtered all the Spaniards and threw their bodies over the ramparts into the ocean and on the beach.

In spite of this absolutely astonishing confirmation of the story that he laboriously reconstructed in his inner exploration, Karl was not satisfied. He continued his library research until he discovered a special document about the battle of Dún an Óir. There he found that a priest had accompanied the Spanish soldiers and was killed together with them. The initials of the name of the priest were identical with those that Karl had seen in his vision of the seal ring and had depicted in one of his drawings.

REVISITING THE SALEM WITCH HUNT

Our observations concerning past life experiences are not limited to our work with clients, workshop participants, and trainees; many of them happened in our own lives.[7] In 1976 Christina and I lived for several months in the Round House at the Esalen Institute in Big Sur, California. It was a small charming structure located by the creek dividing the Esalen property into two parts. Its busy waters rushed down from the mountain ridge and formed a large waterfall just before joining the Pacific Ocean. In front of this house was an opening in the ground spewing hot mineral waters into a little private pool. According to the local lore, the Esalen hot springs had their origin in an interconnected system of volcanic underground caverns that spread under much of California.

The rushing of the creek and the roaring of the waterfall represented a very powerful sensory input. However, even more impressive was the psychic power of this place. Over the years, we have invited to our Esalen workshops as guest faculty many people with extraordinary psychic abilities—clairvoyants, shamans from various parts of the world, members of the Spiritist Church, Indian yogis, and Tibetan masters. They all seemed to concur that the area around the Round House was a "power spot," a place endowed with extraordinary spiritual energy. Those trying to find some scientific explanation for the impact it had on people attributed it to high concentration of negative ions due to the proximity to the ocean, the crashing waters of the waterfall, and to the presence of giant redwoods lining the creek on both sides.

In any case, whatever was the reason for it, living in the Round House had a very powerful psychological impact on both of us. It was unusually easy to go into a meditative state; I found myself often slipping into a trance, in which I forgot our geographic and historical coordinates and felt that our little eyrie was situated somewhere in an archetypal domain beyond time and space. Christina, who was at that time undergoing her spiritual emergency, experienced an extraordinary intensification of her inner process. One weekend, her experiences reached such intensity that they resembled a psychedelic session.

After a period of intense anxiety and uncomfortable physical feelings, she had a powerful experience of what she felt was a memory from one of her past lives. She became an adolescent girl living in New England's Salem who had episodes of nonordinary states of consciousness. Her fundamentalist Christian neighbors interpreted these episodes in their bigoted craze as possession by the devil. This brought an accusation of witchcraft; she was tried by two judges dressed in ceremonial robes and sentenced to death by drowning.

In the middle of her experience, Christina recognized that these two judges were in this lifetime her ex-husband and her father, both of whom were educators. She suddenly understood why she always had a strong negative reaction when she saw them in their dark academic habits performing school ceremonies. It also seemed to her that the difficulties she had had with them in this lifetime were, at least partially, karmic carry-overs from the Salem episode. The past life memory culminated in an experience of execution by drowning. Christina found herself being carried to a pond, tied to a board, and submerged, head first, under water. She managed to notice that the pond was surrounded by birch trees. As she was reliving her death by drowning, she was screaming, choking, and bringing out a lot of mucus, both through her mouth and nose.

While living in Hawaii, Christina had suffered from severe allergies and sinusitis. She had many medical examinations, tests, and treatments, including a series of injections for desensitization. Her doctors, frustrated by the failure of all therapeutic efforts, finally suggested a surgical procedure, involving scraping and cleaning the sinus cavities. Christina decided to refuse such a radical procedure and accepted her predicament. She discovered to her great surprise that, following the episode in which she relived the Salem trial and death, her sinus problems disappeared.

Fortunately, by this time, my belief in what I once had held as a "scientific worldview," which had been rigorously proved beyond any reasonable doubt, had already been seriously undermined by many similar observations. Without it, this episode would have left me in a serious intellectual crisis. There was certainly a certain element of cosmic humor in the fact that Christina's difficulties—which had resisted the concerted efforts of scientific experts—were resolved by reliving a karmic episode involving ignorance, religious fanaticism, and the false accusation of witchcraft.

This episode had a very interesting sequel many years later when Christina and I visited Boston to conduct a holotropic breathwork workshop. The workshop ended in the evening and our flight back to San Francisco was not until late afternoon the next day. We thus had a good part of a day for sightseeing. We decided to call Marilyn Hershenson, a psychologist and dear friend of ours, who had been a member of Swami Muktananda's inner circle. We had become very close in the early 1980s, when we had jointly coordinated a large international transpersonal conference in Bombay. Marilyn was very excited and offered to spend the day with us and drive us around.

Marilyn suggested that we have lunch at her favorite restaurant situated on the ocean near Salem. As we approached it, we found out that its name was Hawthorne Inn. This immediately brought to mind Nathaniel Hawthorne, his *Scarlet Letter,* and the topic of witchcraft. As we were having lunch, Christina recounted for Marilyn the story of her past life memory involving the Salem witch trials. Marilyn was astounded, because she had relived a similar episode of her own in one of her meditations in the Siddha Yoga ashram.

Because we were only several miles from Salem, it suddenly seemed very appropriate to visit this town between lunch and our departure for California. As we were entering Salem, Christina asked Marilyn if there was a pond there. Marilyn, who had spent her entire childhood in Salem, decisively denied it. But then she suddenly took a wrong turn by mistake, which was surprising, since she knew the town well. This unplanned detour unexpectedly brought us to a pond on the shore of the ocean. It seemed like it had originally been a bay, now separated from the body of water by an old stone dam.

Christina got out of the car, as if in a haze. She was looking around and seemed disappointed. "I don't see any birch trees," she said and started walking around the pond. "Where are you going?" we asked her. "There have to be some here," she said and continued walking. We parked the car and followed her. Finally, on the other side of the pond, Christina discovered a birch tree; its trunk was broken and its crown was submerged in water. "You see, they were here," she said. "This must be the last one."

We returned to the car and decided to visit the courthouse where the trials had been held. On the way there, Christina told Marilyn that she had recognized the two judges in her past life experience as her ex-husband and father in her present life. "But there was only one judge at the trial," Marilyn objected. "There were *two judges*!" Christina said assertively. When we came to the courthouse, we found out it was closed. But by the front door was a large tablet, describing the trials. Not only did it confirm that there were two judges, but one of them was called Corwin, which suddenly seemed like a cosmic joke, Christina's ex-husband's name being Win.

Before we returned to the car, I bought from a gift shop a little illustrated booklet on Salem, which included the story of the witchcraft trials. As Marilyn drove us to the airport, I read out selected passages from this booklet. We found out that the girls who had been accused of witchcraft had spent much time with a slave servant named Tituba, who had been accused of being the link to the devil. Tituba was an Indian from an Arawak village in South America, where she had been captured as a child, then taken to Barbados as a captive and sold into slavery. We concluded that Tituba had probably been teaching them some shamanic techniques that had been misunderstood and misinterpreted by ignorant neighbors as work of the devil.

The most interesting information I found in the guidebook was that Old Salem, where many of the historical events had happened, was currently called Danvers. That came as a shock to us. Danvers was the place where we had held a large conference of the International Transpersonal Association (ITA) in 1978. There we had presented for the first time our concept of spiritual emergency, implying that many episodes of nonordinary states of consciousness that mainstream psychiatrists diagnose as psychoses and often treat with drastic methods, such as insulin coma and electroshock, are actually psychospiritual crises.

In our lecture in Danvers, we had suggested that—properly understood and supported—these crises of spiritual opening can actually be healing, transformative, and even evolutionary. We had given the talk in a hall from which we could see, on the other side of the valley, an old-time psychiatric hospital that had one of the worst reputations in the country. They were still using shock methods, which bore great similarity to the practices of the Inquisition and similar witch hunters.

We were stunned by this incredible synchronicity. Of all the possible locations, we presented our modern plea for a radical change of attitude toward nonordinary states of consciousness in a place that, unbeknownst to Christina, had been the site of her past life memory in which her suffering and death had been caused by misunderstanding and misinterpretation of nonordinary states of consciousness.

The existence of past life experiences with all their remarkable characteristics is an unquestionable fact that can be verified by any serious researcher who is sufficiently open-minded and interested to check the evidence. I have given in this article three examples of past life memories experienced by adults in holotropic states of consciousness. Even more compelling and challenging material was amassed by Ian Stevenson, a Canadian psychiatrist who conducted meticulous research with more than three thousand children remembering their past lives; he was able to show the veracity of an astonishing number of these memories.

While all these impressive facts do not necessarily constitute a definitive "proof" that we survive death and reincarnate as the same separate unit of consciousness, or the same individual soul, they represent a formidable conceptual challenge for traditional science, and they have a paradigm-breaking potential. It is clear that there is no plausible explanation for these phenomena within the conceptual framework of mainstream psychiatry and psychology. Having observed hundreds of past life experiences and experienced many of them myself, I have to agree with Chris Bache that "the evidence in this area is so rich and extraordinary that scientists who do not think the problem of reincarnation deserves serious study are either uninformed or "bone-headed."[8]

CONCLUSIONS

My observations indicate that we can obtain information about the universe in two radically different ways. Besides the conventional possibility of learning through sensory perception, and analysis and synthesis of the data, we can also find out about various aspects of the world by direct identification with them in an altered state of consciousness. Each of us thus seems to be a microcosm that also has access to the information about the entire macrocosm. This situation is reminiscent of the descriptions found in the ancient Indian spiritual systems, particularly in Jainism and in Avatamsaka Buddhism.

According to Jain cosmology, the world of creation is an infinitely complex system of deluded units of consciousness, or *jivas*, trapped in different aspects and stages of the cosmic process. Each jiva, in spite of its seeming separateness, remains connected with all the other jivas and contains knowledge about all of them. The *Avatamsaka Sutra* uses a poetic image to illustrate the interconnectedness of all things. It's the famous necklace of the Vedic god Indra: "In the heaven of Indra, there is said to be a network of pearls, so arranged that if you look at one, you see all the others reflected in it. In the same way, each object in the world is not merely itself, but involves every other object and, in fact, is everything else." Similar concepts can be found in the Hwa Yen school of Buddhist thought, the Chinese version of the same teaching.

Gottfried Wilhelm von Leibniz's monads—ultimate elements of the universe—resemble the Jain jivas, since each of them contains the information about all the others.[9] In the esoteric traditions, this was expressed by such phrases as: "as above so below" or "as without, so within." In the past, this basic tenet of esoteric schools, such as Tantra, the Hermetic tradition, Gnosticism, and Kabbala, appeared to be an absurd confusion of the relationship between the part and the whole and a violation of Aristotelian logic. In the second half of the twentieth century, however, this claim received unexpected scientific support by the discovery of the principles operating in optical holography.[10]

Ancestral, racial, collective, karmic memories; experiential identification with animals, plants, and inorganic materials and processes; and other types of transpersonal experiences meet the criteria for "anomalous phenomena." Their existence and nature violates some of the most basic assumptions of materialistic science. They imply such seemingly absurd notions as the arbitrary nature of all physical boundaries, nonlocal connections in the universe, communication through unknown means and channels, memory without a material substrate, nonlinearity of time, and consciousness associated with all living organisms and even inorganic matter. Many transpersonal experiences involve events from the microcosm and the macrocosm, realms that cannot normally be reached by unaided human senses, or from historical periods that precede the origin of the solar system, the formation of planet Earth, the appearance of living organisms, the development of the nervous system, and the emergence of *Homo sapiens*.

I do not know any scientific conceptual framework other than Ervin Laszlo's connectivity hypothesis and his concept of the Akashic field that would provide a reasonable explanation for these paradoxical findings. The observations from research on holotropic states of consciousness suggest that the individual human psyche has—at least potentially—the capacity of accessing all the information stored in the Akashic field. It can do it not only in the role of an uninvolved observer but also as a protagonist, identifying experientially with all the people, animals, plants, and even inorganic objects and processes that are part of this field.

I have been so far talking about phenomena that belong to the material world. One fascinating question remains wide open for further exploration. Laszlo's Akashic field is an excellent model for the historical aspect of C. G. Jung's collective unconscious.[11] However, Jung's collective unconscious also encompasses another domain, which is radically different from the former—the archetypal realm with its mythological beings and abodes of the Beyond. And yet, in holotropic altered states, visits to archetypal regions and encounters with archetypal figures often appear on the same continuum and in various combinations with experiences reflecting phenomena from the material world. This suggests that—in some yet unexplained way—traces of the archetypal world are part of the Akashic field, and the Akashic field also mediates experiential access to this domain.

EIGHTEEN

Dialogues with My Dead Brother

Fr. François Brune

François Brune was ordained as a priest of the Roman Catholic Church in 1960. He studied philosophy and theology in Paris, Paderborn, and Tübingen and holy scripture in Rome. He is a graduate in theology of the Institut Catholique in Paris and in the field of Holy Scripture at the Istituto Biblico in Rome. He was professor of theology and holy scripture in several seminaries in France. He has written twelve books translated into seven languages and delivered lectures in various countries in Europe and North and South America.

Since the dawn of humankind, there have been people in every civilization who have had communication with the dead through mediums. But as we have developed new technical possibilities it seems that the dead have been trying to use them for better forms of communication. In some cases a living person, usually in the presence of a medium, has a telephone conversation with a dead person. It's difficult to know precisely where and with whom this development started. The first case I know of was in Brazil, between 1917 and 1925, which involved telephone communication mediated by a man named Oscar d'Argonnel.

Instances of after-death phone communication have since been reported in various countries all over the world, but they remain rare.

After-death contact has also been achieved through TV screens and computers, and more often through tape recorders. These ways of communicating "with the beyond" through electronic devices are called Instrumental TransCommunication or ITC. (Some speak of Electronic Voice Phenomena or EVP, but that applies only to voice communication whereas images and text can also be received.) The various forms of ITC were developed initially in Germany and Italy.

After many years of involvement in this field, I believe that I have met all the major ITC researchers in the world. I have been with them on numerous occasions to observe how they proceed but did not attempt communication myself; that's not my task. My task is to study the results, obtain the documentation, and publish it. The same held true for the experiences I am going to share with you here in which contact was made with my deceased brother by two of my friends, but it was made in my presence.

TRANSCOMMUNICATION WITH MY BROTHER

My elder brother died on April 24, 2006, in Paris. On May 27 I was in Caen, a town in Normandy, to give a talk at a meeting about death, life after death, and communication with the beyond. The next day, on May 28, exactly one month and four days after the death of my brother, I was in the apartment of my younger sister in town with some of the guest lecturers. The contact with my brother was attempted by two friends who had come to the conference from the city of Toulouse. They knew nothing about my family; they only knew that Jean-Pierre Brune was my brother and that he had died.

One of my friends sat in the bedroom trying to receive news from my brother through what is known as automatic writing. The other friend was with me in the dining room, with a medium, my younger sister, and some others. (The others had no actual function—they were with us only to witness what we are doing and perhaps help with their love and prayers.) In the dining room we were trying to obtain information from my brother through a tape recorder. We were separated from the bedroom only by a door that was sometimes open and sometimes closed. On each side of the door stood a cameraman to film the experiment.

The setup was very simple: We had a small radio tuned to broadcast first in Dutch, and then in German. (It appears that it's easier for the deceased to transform existing sound waves than to create them from zero.) About forty centimeters in front of the radio we had a small tape-recorder with a device that could make it run faster or slower. This is necessary because the voices often come too fast to permit ready understanding. When we entered into contact with the dead, the tape recorder recorded our questions with a small pause between them to give those on the other side an opportunity to answer. During the time that we did not hear anything "from the beyond," the tape recorder recorded the broadcast from the radio.

Then, about fifteen or twenty minutes after the contact had ended, we rewound the tape to the beginning and listened. We heard the broadcast, and our questions, but sometimes also a voice that seemed to have come from the beyond. At times the sound of the broadcast disappeared completely and transformed into a new voice, while at other times it was possible to hear the voice from the radio with another voice superimposed on it, which had no relation to the broadcast but was meaningful in regard to our questions.

It was soon clear to us in the dining room that communication had begun in the bedroom: we could see that the pencil was moving on the paper. But when we wanted to begin the tape recording, the cameraman on our side asked us to wait a little: his equipment was not able to receive any sound. He went back to his car for other equipment and replaced something on his camera, but without success. He went back again to his car and substituted further equipment, but still without success. He was very disappointed and explained that filming had been his job for the past ten years; he had the best equipment available and never had such a problem. He gave up completely and laid his equipment on the table. Having merely the image of people speaking in the room without hearing what they said would not have been of interest.

Nevertheless, we got the tape started and began to speak to my brother, leaving a moment of silence after every few sentences. After we had been speaking for some time, the cameraman, who had not been handling his apparatus, suddenly exclaimed in astonishment: "Now it works!" He immediately began to film the rest of the experiment.

After we had posed some questions to my brother, the friend from the bedroom came over to join us at the tape recorder. We finished one side of the tape, rewound it to the beginning and began to hear and note what we had received. It was only after this that we asked our friend in the bedroom to show what he had written in automatic writing.

In order to understand the messages he received in this form of writing I should explain that my brother's wife had died many years previously and that they had three daughters. The eldest is an actress at the Comédie Française, the second studies moviemaking at the famous Lumière school, and the third is working for a cinema and theater casting company.

Here is the translation of what my friend received. The words were clear, although the style of the writing was not always correct and the words themselves were not always the right ones.

The appointment was well made, and what a surprise to converse so soon! As I arrived, mother was there, faithful. She was very different from before. I recognized her immediately. She was more beautiful, younger. She looked at me as she did when we were children, protectively. She didn't move her lips, but I heard her speaking to me and I recognized her voice. She said "my blond angel." That brought me automatically back to the past. I was smaller by a head; nothing had changed. She said to me that now nothing could keep me from breathing.

I am learning very quickly here, but you must be patient. I am not yet able to be in two different places at the same time. I remain at first with you here, and I delay the séance to act fully here. Your friends are very kind. Do you hear this melody? It will be played for you by my friends, who are all piano players, without the key of G. I am playing thousands of pieces of music in the heart, above all in the hearts of my children. I help them, too, because sometimes they are living very painful moments; but I don't know yet what to do to help them. Mum wants to help me; she wishes them to be artists. She prepared for them hundreds of stars. In one month I had hoped to do very much, but I cannot yet do everything. I get down to work for the daughters, as well as for the theatre. I am blocking the sound, I will have to go there . . .

Tell François his thoughts are forces and energies very useful for those who, like me, are discovering the endless horizons. The Father [the priest], the friend of François, asks him to remember Paolo. Mum loves all of you.

The sound does not get through; I must go there and then come back.

At this point my friend's hand didn't receive any impetus for movement for a few seconds, and it was at this very moment that the equipment of the cameraman began to work in the dining room. My brother's reference to Paolo was very pertinent, as he is an Italian aeronautics engineer who specializes in high frequencies. He works with Daniele Gulla in the Bologna laboratory of ITC. I have known him for many years and had met him again some weeks earlier in Vigo, Spain, at an international conference on ITC where we were both lecturers. I had never spoken of him either to my sister or to the friends from Toulouse who were making the contact with my brother.

The automatic writing began again:

"Put piano [music] as background."

The friend who was doing the writing stood up and went to the dining room to arrange this. My sister immediately put on a tape of piano music. My brother's reaction was:

"It's better than before, the vibrations of music are nearer to ours, they can help to find me."

Then came a written answer to a question we had asked on the tape.

"Yes, it's me, and by the painting it will be more magic. Yes, pray for me, I am alive, don't forget Paolo."

My brother had told us that his new friends in the beyond would play a melody for us. In the dining room, only one of us, the medium, heard this music. My friends didn't know that my brother was blond, that he played the piano very well, and had his own piano in his bedroom until the end of his life.

Compared to what we received through automatic writing, the content of our reception via the tape recorder seemed poor: there were only a few words. Nevertheless, the words were important because they were direct answers to questions we had posed. It is also interesting to note the agreement between the messages received by automatic writing and by tape.

Some of the messages we received were hard to hear and others were very clear. Sometimes it seemed that it was my brother who was answering, sometimes someone else. And at times the answer didn't correspond to the question. I give here only the clearest and most important instances.

To our first question my brother answered only: *"I have come."* Some seconds later, a female voice said: *"He has breathed"* and a little later, *"He is free."* At the end of his life my brother had difficulty breathing, and through the automatic writing our mother said that nothing would now keep him from breathing.

I then asked: "If my brother cannot speak, perhaps somebody else could give us some news from him?" Four or five seconds later a female voice said: *"Love."* Then I said, "We are in the apartment of our sister in Caen; you too came here often." Some ten seconds later a female voice confirmed, *"He came."*

My brother was very fond of whiskey. At the end of his life he knew he had not much time left to live, and he drank still more. So I asked him: "Do you want a glass of whiskey to make the conversation easier?" He answered in a joking tone: *"I accept whiskey."* These words were particularly clear. (As we shall see, the story about whiskey continued in the second and third contact.)

One of my friends asked my brother if he had something to tell me. But a female voice answered: *"I loved François."* The last question: "Do you want to have a new call?" and the answer: *"Yes, I am waiting, yes."*

SECOND CONTACT WITH MY BROTHER

The second contact was made on January 28, 2007, again following a conference in Caen. It was organized a little differently. The writing séance came first and was not automatic but telepathic, that is, my friend could hear the words in his head and then write them down in his own hand.

For the taping session two of the three daughters of my brother came, as well as some friends, one of them a doctor specializing in anesthesia and interested in near-death experiences. Altogether fourteen people were present.

I give here only the most important elements, leaving out passages that were too personal. Word for word, the message received by the friend who listened and wrote began:

> *I have heard you calling for a new meeting. I feel myself much lighter, fluid. Learning is as varied as possible; knowledge is universal.*
>
> *Now I know the deep meaning of spirituality. What I could not reach on Earth is now my strength here. In effect, François, now I have understood. I am able now to understand your life and your devoutness, which before was impossible for me to understand . . .*
>
> *I listened to you yesterday, to you also François, I have heard everything, your researches about mediumship and saintliness. I'll be with you also this evening in that place full of intense vibrations. There, too, saintliness and mediumship were united. . . .*

Then came a long paragraph about Mont Saint Michel, a sacred and famous site in France. My brother's younger daughter burst out laughing. "Of course," she said, "Dad was a teacher of geography and history! He is giving lessons again." (I should add that my friends from Toulouse who were making the contact didn't know that I had decided to go that afternoon with two others to Mont Saint Michel for the night.)

> *Yes, my daughters, thanks for your efforts, your love. You ought to recognize when I am there; I can become manifest very easily through music and electricity. You know it!*

This refers to the fact that in the apartment of my nieces abnormal phenomena involving lights and loudspeakers were taking place from time to time. And then, at the end came these words: *"I'll join you in drinking a* coupe *even though some are missing."* (I should explain that in France the word *coupe* is used only for the glasses used for drinking champagne; the word for the glasses for ordinary wine is *verre*. While wineglasses can have different shapes, the glasses that have the particular shape of a coupe [or *flûte*] are used only for champagne.) We were rather astonished, because champagne was not "on our program" for this contact. But then two of the friends we had invited pulled two bottles of champagne out of a sack. We didn't know that they had brought them for us, but my brother did. Of course, my sister didn't have fourteen "coupes" at hand. So, some of us drank champagne in ordinary glasses.

During the second contact we received fifteen messages via the tape recorder. Here are some of them.

Question: "We are calling Jean-Pierre Brune. Jean-Pierre are you pleased that we are calling you again? Did you expect this moment?"

"I am waiting for you."

Question: "Jean-Pierre, can you tell us something about your evolution in the beyond? Are you authorized to tell us?"

"There is somebody who teaches me to love you . . . it's somebody I like . . ."

One daughter asked: "Dad, did you find Mum?"

"We are together and thanks, thanks, it's beautiful!"

Question: "Jean-Pierre, are you still missing whiskey?" (This time this was not asked by me but by one of my friends.)

"*La bouteille!*" [the bottle!]

THE THIRD CONTACT

The third contact with my brother developed in two stages. A conference was again organized in Caen in March 2008. My friends from Toulouse arrived at their hotel the day before, when one of them felt that he would receive a message from the beyond through telepathically transmitted writing. He immediately took a pencil and paper, and a message from my brother did come:

> *There would be so many things to say! We don't know with what subject we could begin. You are all together in that town, where history has left its mark. This town had to learn how to rebuild itself again. François is always deeply impressed by these roofs, walls, the fortress, the history with the Germans, so much suffering . . .*

Then some paragraphs followed that are too personal to be included.

> *Of course, we will be there with you during these days, as at the other times. You must know that when you are giving, sharing, you give a meaning to your hardships and so others will utilize your testimonies to get further ahead. This story is yours, but it's also of your brethren.*

I believe it's important to note that during this contact my brother was not alone: he was saying "we." We also received voice messages from other persons, both male and female, who were with him in the beyond during this contact.

The conference in Caen took place on March 29, and the next morning we gathered again in my sister's apartment to try to communicate with my brother through tape. Even before we could pose the first question we received two sentences, although they didn't come directly from my brother.

"He is alive . . . you are waiting."

"He is all right."

Question: "Has something changed in your life since our last contact?" Three answers came with just a small pause between.

"I love you more and I am here."

"It's the perfect world that's speaking to you."

"I feel well."

My sister asked: "Are you still with Mum?"

"Very close to my mother . . . I love her."

A question from one of our friends: "I feel you have something to say to your daughter."

"She is alone . . . help her with pleasure. . . . Oh, I understand well, we will not need ten years."

Question: "Are you happy when we call you?"

"It's doing me good."

And then, without waiting for a question:

"I had ordered whiskey." (It seems that there is a sense of humor in the beyond!)

I believe these notes convey the most important elements of our contacts. It's clear to me that my brother began a spiritual evolution, progressing faster than we generally do on Earth. He said this himself at the beginning of our second contact and at the end of the third. He spurred us to communicate these messages to others, to our "brethren." It appears that these messages were given to us not only for our family and friends, but for all humankind.

An evolution of this kind is completely new for my brother. I knew that it was possible, but I dared not hope that it would come about so quickly.

It's also clear that my brother was reacting to what we are doing on Earth. He was trying to interact with our life, but evidently this is not easy. He is alive, but in another dimension, a dimension that may be accessible to us through the Akashic field.

Part Four

REVIEWING AND ASSESSING THE EXPERIENCE

NINETEEN

Epiphany in Space and on Earth

REFLECTIONS ON THE AKASHIC EXPERIENCE

Edgar Mitchell

Edgar D. Mitchell was a member of the Apollo 14 moon mission and was the sixth man to walk on the moon. He holds, together with Alan Shepard, the record for the longest moonwalking session (9 hours and 17 minutes). He is the founder of the Institute of Noetic Sciences, cofounder of the Association of Space Explorers, and author of several books and more than 100 articles and papers in scientific and popular journals. Mitchell received a B.S. in Industrial Management from Carnegie Mellon University, a B.S. in Aeronautical Engineering from the U.S. Naval Postgraduate School, and a Sc.D. in Aeronautics and Astronautics from the Massachusetts Institute of Technology. He has four honorary doctorates and was nominated for the Nobel Peace Prize in 2005.

In February 1971 during the return flight of Apollo 14, following exploration of the lunar surface, my often-described life-changing epiphany occurred. It has taken many years of deep study on my part to find a coherent scientific framework to explain the profundity of this event, while enjoying its emotional, esthetic, and professional satisfactions. What was it about seeing our home planet from a great distance that caused my mind to make a major shift in perceived values and led me to redirect the course of my life into more esoteric pursuits? What is it in nature that stimulated this sense of wonder, awe, excitement, and ecstasy at the most profound levels? The noted British astronomer, Fred Hoyle, predicted at the beginning of the space age that pictures of Earth from space would create major changes in human perception about ourselves. To a certain extent this has happened, as the pictures of our planet from deep space have been continuously in demand, published in both print and electronic media since the first photographs were taken from the vicinity of the moon on Apollo 8 in 1968.

The word *agape* comes to mind, both in meaning a sense of astonishment and within the Greek concept of an asexual love of all things in nature. Subsequent inquiry after my return revealed the fact that such profoundly transformative events have been occurring throughout history in all cultures to some individuals, mostly to certain tribal and spiritual leaders, though not with great frequency in the general population. The ancient Sanskrit phrase *savikalpa samadhi* nicely describes my experience of seeing the separateness and individuality of physical objects, like stars, planets, and galaxies, with my eyes, but experiencing at a visceral level the feeling of connectedness, or unity, of all matter born in the furnaces of star systems, including our biomolecules. The experience was accompanied by a sense of ecstasy and bliss.

All who have traveled into space and looked back at this magnificent blue and white haven of life have wondered and marveled at the implications of living things existing on an isolated little planet, nestled around an average mainstream star, which is one of billions in our galaxy. How many other life-bearing planets exist around those stars? The reactions of we early space farers has been called the "overview effect" in a book by Frank White (1998). The author, and all of us referenced in the book, will agree that if more humans could have the opportunity to view our planet from afar, life on Earth would be significantly transformed for the better.

If this idea is true, why is it true? What is it that causes us humans to renounce a strong ego-driven self-interest, even greed, in order to pursue an idea, a dream, a goal that may or may not be realized but which offers the hope for a more peaceful and harmonious world for all people? All space farers during the first fifty years of human space flight have remarked on a world without obvious boundaries, one that appears harmonious and inviting below the white clouds and blue tinge of the surrounding atmosphere. That appearance belies our knowledge of the conflict and strife historically existing between cultures and nations.

FROM COSMOS TO QUANTUM

In my case, I was drawn to the idea that humans of all epochs have looked to the heavens and pondered the questions of our origins and our relationship to the larger cosmos. I realized that the cosmology from modern science was certainly incomplete and might be flawed and that the cosmologies from our cultural beliefs, rooted in ancient local religions, were archaic and likely flawed as well. A most fundamental difference between science and the cultural cosmologies is that science avoided the issue of "what is consciousness" during the four hundred years after philosopher René Descartes quieted the Inquisition by declaring that body and mind belonged to different realms of reality, thus allowing science to arise as a materialist pursuit in the Western world apart from religious influence and persecution. As a result, most science has assumed during those four centuries a reductionist approach and considered consciousness to be an epiphenomenon of molecular complexity, with evolution itself considered a random process of nature. Fortunately for us in the modern period, we have been able to penetrate the error of Cartesian dualism (separateness of mind and body) and find in the discipline of quantum physics some new answers that help resolve these difficult questions about mind-matter interactions.

The ancient Hindu concept of Akasha as the fundamental of the cosmos, an ether that contained both material and nonmaterial substances, led to the notion in metaphysics that experience in nature was not lost, but preserved. This was labeled the *Akashic Record,* though how preservation of experience might be accomplished was not specified. Science produced no corresponding concept, nor a mechanism to validate one, until late in the twentieth century with the work of Ervin Laszlo and the discovery of the quantum hologram by Professor Walter Schempp. Schempp—while investigating how to improve the process of magnetic resonance imaging (MRI), a topic quite unrelated to the present issue, or so it was originally thought—has opened new vistas in our understanding. As Laszlo describes in his Summing Up, the quantum hologram is a nonlocal quantum information structure derived from Max Planck's study in the late nineteenth century of the surprising radiation emitted by material substances, called "black body radiation." For most of the twentieth century such radiation was believed to be curious random photon emissions from matter, and of minimal interest—until Schempp demonstrated that the emissions are entangled, coherent, and carry nonlocal information about the emitting object. Subsequent studies have shown such nonlocal information to be fundamental not only to our normal perceptual faculties but also that it forms the basis of intuitive-level information.

In the English language *intuition* is referred to as the "sixth sense." However, intuition is now known to be based on quantum information—which existed even before our solar system and planet were formed to provide a basis for the five normal senses—so intuition should really be labeled as our "first sense."

RELEVANCE TO HUMAN LIFE AND EXISTENCE

One may reasonably ask: "What has this to do with sudden changes in our worldview, and particularly a change in our value system toward emphasis on unity and service to the greater public good?" Two converging lines of approach to the question should be considered, both rooted in ancient issues of the origin and nature of existence.

First of all, on a grand scale of human existence, clearly unity, peace, and cooperation are values and objectives that should be more conducive to the survival of the species than conflict and violence. At least, intellect would suggest that's so. However, conflict and resulting violence has existed and been rationalized throughout human history, and certainly in prehistory as well, where tribes battled other tribes for territory, food, mates, and material abundance. And the food chain, where some species eat other species, and cannibalism, where certain species, including humans, dine or have dined upon their own, are historic fact. These issues of existence and behaviors in nature go to the heart of the forces that shape our worldviews and value systems on Earth.

Second, and of equal importance, at the base of all evolved thought systems are concepts of transcendence, where one experiences an insight so moving that old values that permit and rationalize violence to self and others become unthinkable. Our question pertains to the cause of transcendence and transformation. Need we rely upon traditional religious explanations involving deities? Or is there a more natural mechanism we can discover?

The discipline of meditation, in a number of related forms, helps one to learn to quiet and control the mind's restless roving and to achieve a calm and focused state. It has been a method for centuries for learning to manage and control unwanted behaviors and to achieve a state of peace and happiness within. And certainly, creative faculties flourish better in the soil of a rested, focused, and disciplined mind. Transcendence and transformation of consciousness in individuals within religious orders following such meditative practices have been noted for centuries. Avatars and sages arise from such practices and have long been admired and followed for their wisdom and service to the greater good of humankind.

Cloistered orders devoted to the service of humankind and nature usually derive their authority and direction from a belief in a deity or deities and from a long line of traditional practices. Although there is certainly a long history of warfare and violence attributed to the political influence of certain religious beliefs, the inner core of cloistered life is more often centered on practices devoted to peace and service to the greater good of individuals and the community. Insights and epiphanies regarding service to others are not unusual among such practitioners and are expected as part of the discipline. However, that does not explain why individuals who are not in religious orders or engaging in religious practices have spontaneous events that lead to insights and similar behavior modifications that are similar to what happened for me when I saw the expanded views of the heavens from the vantage point of space.

An answer may be found in the Akashic Record, more exactly in the Akashic field, augmented continuously by quantum holographic information. To routinely access this deeper level of intuitive information requires a natural openness to such information, enhanced by practice, and by learning to trust the validity of such experience. As suggested above, intuition should be considered a basic source of information (our "first sense") available in nature long before humans evolved to use language and so-called left brain processes. Intuitive information affects us at the cellular level and is more associated with feeling than with thinking, intellect, and language.

It's now quite clear from studies that animals use this quantum-entangled information quite routinely in a prelinguistic manner. Note, for example, the coordinated movements of flights of birds or schools of fish as they move collectively through their environment. Similarly, well-trained athletic teams seem to function as a single unit when "in the zone" and emotionally connected in this manner. When we humans function as individual intellectual organisms, this feeling-level of information and insight is suppressed. Yet, as the reports published in this volume show, similar worldview and personal behavior changes occur in near-death and out-of-body experiences as well.

Regarding major and sudden changes in beliefs, values, and behaviors, I postulate that exposure of our mind to an experience or venue that's exceptionally astonishing or breathtakingly beautiful, such as the classic mountaintop experience or the experience of seeing Earth from deep space, opens the brain to quantum-level information. Throughout history the impact of such Akashic experiences on the individual's emotions has proved to be largely irreversible. The sense of awe, peace, well-being, and bliss that accompanies such experiences simply cannot accept violence, conflict, and disharmony as viable behavior options at the personal, social, or cultural levels. Whatever accommodation the mind had previously made to rationalize, justify, or overlook such behaviors is no longer possible.

It's challenging to devise a protocol within the scientific method to test, validate, or falsify this hypothesis. However, whether or not we can validate the proposed cause of these life-changing Akashic experiences with the protocols of science, it's likely that they will continue to occur as we humans find ourselves exploring new venues in nature that inspire awe and wonderment and expand our perspective.

TWENTY

Nonlocal Mind, Healing, and the Akashic Phenomenon

Larry Dossey

Larry Dossey is a physician of internal medicine and former chief of staff of Medical City Dallas Hospital. He is the author of eleven books dealing with consciousness, spirituality, and healing. Dossey is the former cochairman of the panel on Mind/Body Interventions, National Center for Complementary and Alternative Medicine, National Institutes of Health. Dr. Dossey is the executive editor of the peer-reviewed journal *Explore: The Journal of Science and Healing*. He lectures around the world.

The Akashic experience, as Ervin Laszlo explains in this book, reflects the wider dimensions of reality. This experience should not be considered as merely a metaphorical or poetic way of speaking but as a universal aspect of human experience.

THE TWO FACES OF NONLOCAL MIND

The Akashic experience reflects what I've called *nonlocal mind*—mind that's infinite in space and time, unconfined and unconfinable to the here-and-now.[1]

As a physician, my attention has naturally been drawn to ways in which these experiences relate to human health. Thus I have found it useful to regard these experiences as being of two types. In one, we *acquire* information from the world, and in the other, we *insert* information into it. Both these categories transcend the physical senses and are inexplicable by conventional, classical science. In both these situations, the customary barriers of space and time fade, and we manifest as nonlocal or infinite beings whose consciousness operates in ways that are not limited to the here and now.

As the Hindu sages who fashioned the Akashic concept realized, there are aspects of the human mind that are unlimited in space, therefore omnipresent, and that are also boundless in time, therefore eternal and immortal. Omnipresence and eternality are qualities that have always been attributed to the Divine—thus the Hindu aphorism *Tat tvam asi,* "Thou art that," which affirms that we share qualities with the Godhead or the Absolute, however named.[2]

One of the most common ways we acquire information nonlocally is in precognition or premonitions, literally "a knowing that comes before." Surveys show that the great majority of individuals experience premonitions, most commonly in dreams. Recent computer-based, laboratory experiments by researchers Dean Radin, Dick Bierman, and others show that the ability to sense the future may be intrinsic to some degree in most individuals. In these stringently controlled studies, individuals respond to certain kinds of images physiologically and unconsciously a few seconds before the computer randomly selects and displays them. This innate ability is called presentiment, "a feeling that comes before."[3, 4] These experiments are of the highest importance, because they demonstrate decisively that human knowing is not time-bound and limited to the present.

The ability to acquire information nonlocally is not merely a laboratory curiosity but has been put to practical use. Archaeologist Stephan A. Schwartz, a founder of the field known as "remote viewing," has used this technique repeatedly to find sunken ships and buried archaeological sites that have been lost to history.[5] The odds against a chance explanation for these discoveries is simply staggering and is evidence that nonlocal knowing can confer practical benefits to those who have the courage to claim it and put it to use in their life. In fact, nonlocal knowing has been successfully employed for decades in the field of archaeology. This little-known history is described by Schwartz in his fascinating book *The Secret Vaults of Time.*[6]

The ability for nonlocal mental functioning may have become encoded in our genes in our evolutionary ascent, because this capacity would be likely to contribute to the endurance of the individual possessing it. Knowing ahead of time where danger lay, or where to find food or shelter, would be of obvious benefit in the high-stakes game of survival. Sensing the future apparently continues to function in this way, even in modern life. In *The Power of Premonitions,* I describe many examples in which people sense impending disasters, avoid them, and survive.[7]

A large body of information suggests that we can nonlocally acquire information about others who are beyond sensory contact. Hundreds of these experiences have been documented. The individuals involved are usually emotionally close—spouses, parent-child, siblings, or lovers.[8] British author David Lorimer calls these connections "empathic resonance," emphasizing the intimate feelings that underlie them. Researcher Guy Playfair has documented the frequent occurrence of these phenomena in twins.[9] Psi researcher Dean Radin considers the experiences evidence of "entangled minds," referring to the phenomenon of quantum entanglement that, he hypothesizes (and Laszlo confirms below), provides a possible explanation for them.[10]

Trans-spatial connections between distant individuals have been demonstrated by studies examining correlated brain functions of distant individuals. In brief, when the brain of one individual is stimulated in some way, the brain of a distant, paired individual demonstrates the same change. The two distant individuals are often emotionally close. These correlations have been shown in experiments involving the electroencephalogram (EEG), which measures electrical activity in the brain, and functional magnetic resonance imaging (fMRI), which indicates metabolic activity in the brain.[11] The flip side of the nonlocal *acquisition* of information is the nonlocal *insertion* of information. As the remote healing experiments reported in this volume show, we can insert information remotely, not only in space but also in time. Radin has reviewed dozens of experiments suggesting retrocausality, the "backward influence" of intentions on events that lie in the past, which we presume have already happened but which retain the ability to be modified under certain conditions.[12]

The possibility that our mental intentions might exert an influence outside the here and now is generally considered to be scientifically blasphemous. As Radin writes:

> These implications [of nonlocal mental actions], of course, are heresies of the first order. But I believe that if the scientific evidence continues to compound, then the accusation of heresy is an inescapable conclusion that we will eventually have to face. I also believe that the implications of all this are sufficiently remote from engrained ways of thinking that the first reaction to this work will be confidence that it's wrong. The second reaction will be horror that it may be right. The third will be reassurance that it's obvious.[13]

HEALING WITH INTENTIONS

One of the most ancient ways in which humans attempt to insert information nonlocally into the world is through healing intentions. When this effort involves a spiritual or religious context, it's often called prayer.

The idea that prayer can affect living organisms is a universal belief spanning ideology, religion, culture, and race and has endured for at least fifty thousand years. As Schwartz states:

> The shamanic cave art of Altamira, Tres Frères, and Lascaux presents compelling testimony that our genetic forebears had a complex view of spiritual and physical renewal, one that has survived

> to the present unchanged in at least one fundamental respect. The intent to heal, either oneself or another, whether expressed as God, a force, an energy, or one of many gods, has consistently been believed to be capable of producing a therapeutic result.[14]

But just what is spirituality? I consider it a felt sense of connectedness with "something higher," a presence that transcends the individual sense of self. I distinguish spirituality from religion, which is a codified system of beliefs, practices, and behaviors that usually take place in a community of like-minded believers. Religion may or may not include a sense of the spiritual, and spiritual individuals may or may not be religious. I regard prayer as communication with the Absolute, however it's named, and no matter what form this communication may take. Prayer may or may not be addressed to a Supreme Being. Many forms of Buddhism, for instance, are not theistic, yet prayer, addressed to the universe, is a vital part of the Buddhist tradition.

Healing Research

Does prayer work in an empirical sense? Rudolf Otto, the eminent theologian and scholar of comparative religions, asserted that it's "a fundamental conviction of all religions" that "the holy" intervenes "actively in the phenomenal world."[15] This is an empirical claim, and science is the most widely accepted method of adjudicating such claims.

The earliest modern attempt to test prayer's efficacy was Sir Francis Galton's innovative but flawed survey in 1872.[16] The field languished until the 1960s, when several researchers began clinical and laboratory studies designed to answer two fundamental questions: (1) Do the prayerful, compassionate, healing intentions of humans affect biological functions in remote individuals who may be unaware of these efforts? (2) Can these effects be demonstrated in nonhuman processes, such as microbial growth, specific biochemical reactions, or the function of inanimate objects?

The answer to both questions appears to be yes. In a 2003 analysis, Jonas and Crawford found:

> over 2200 published reports, including books, articles, dissertations, abstracts and other writings on spiritual healing, energy medicine, and mental intention effects. This included 122 laboratory studies, 80 randomized controlled trials, 128 summaries or reviews, 95 reports of observational studies and nonrandomized trials, 271 descriptive studies, case reports, and surveys, 1286 other writings including opinions, claims, anecdotes, letters to editors, commentaries, critiques and meeting reports, and 259 selected books.[17]

How good are the clinical and laboratory studies? Using strict CONSORT criteria (Consolidated Standards of Reporting Trials), Jonas and Crawford gave an A, the highest possible grade, to studies involving the effects of intentions on inanimate objects such as sophisticated random number generators.[18] They gave a B to the intercessory prayer studies involving humans, as well as to laboratory experiments involving nonhumans such as plants, cells, and animals. Religion-and-health studies, which assess the impact of religious behaviors such as church attendance on health, were graded D, because nearly all of them are observational studies with no high-quality randomized controlled trials.

The depth and breadth of healing research remains little known among health care professionals, including many of those who have offered critiques and analyses of it. Unfortunately, these critiques are almost never comprehensive and are woefully uninformed. Critics typically identify one or two studies that are problematic, ignore the rest, and generalize to condemn the entire field. Or critics may rely on philosophical and theological propositions about whether remote healing and prayer ought to work or not, and whether prayer experiments are heretical or blasphemous.[19] Are these studies legitimate? Should they be done? Dossey and Hufford examined these questions and critiqued the twenty most common criticisms directed toward this field.[20]

It's true that healing research is immature, and anyone hoping to find perfect studies will have to go elsewhere. (Truth be told, perfect studies may not exist in any area of medical science.) Yet this field has already matured greatly and can be expected to continue doing so.

Why do these studies evoke such sharp criticism? It's an article of faith in most scientific circles that human consciousness is derived from the brain and that its effects are confined to the brain and body of an individual. Accordingly, it's widely assumed that conscious intentions cannot act remotely in space and time. The above healing studies call this assumption into question—and this challenge, I suspect, underlies much of the visceral response this field evokes.

What do we really know about the origins and nature of consciousness? As philosopher Jerry Fodor says, "Nobody has the slightest idea how anything material could be conscious. Nobody even knows what it would be like to have the slightest idea about how anything material could be conscious. So much for the philosophy of consciousness."[21] And philosopher John Searle states, "At the present state of the investigation of consciousness we don't know how it works and we need to try all kinds of different ideas."[22]

Are prayer-and-healing studies blasphemous? These experiments are not an attempt to prove or test God, as many critics charge. Above all, these studies are explorations of the nature of consciousness. In view of our appalling ignorance on this subject, it would seem prudent that these investigations go forward, for they might fill in some of the massive blank spots on the current scientific map.

Another frequent criticism of these studies is that they are so theoretically implausible that they should not be done. In other words, they radically violate the accepted canons of science and the known laws of consciousness, and this places them so completely off the scientific map that they don't deserve consideration. Yet, there are no inviolable laws of consciousness. As Sir John Maddox, the former editor of *Nature,* has said, "What consciousness consists of . . . is . . . a puzzle. Despite the marvelous successes of neuroscience in the past century . . . we seem as far from understanding cognitive process as we were a century ago."[23] These studies violate not laws of consciousness but, it often seems, deep-seated, largely unconscious prejudices.

Another common criticism is that these studies are metaphysical; they invoke a transcendent agency or higher power, which places them outside the domain of empirical science. This is a straw-man argument, because researchers in this field make no assertions about entelechies, gods, or metaphysical agents in interpreting their findings. They are searching for correlations between intentions and observable effects in the world. Nearly always they defer on the question of mechanism, which is an accepted strategy within science. Harris et al., for example, in their 1999 study of prayer in patients with coronary heart disease, concluded,

> We have not proven that God answers prayers or even that God exists. . . . All we have observed is that when individuals outside the hospital speak (or think) the first names of hospitalized patients with an attitude of prayer, the latter appear to have a "better" CCU experience.[24]

The Spiritual Factor in Healing

Should physicians concern themselves with the spiritual lives of their patients? Should they pray for them? These questions are unanswerable without first becoming aware of the data in this field. What are the correlations between prayer and other religious behaviors, and health and longevity? What is the effect size? What about risk, cost, availability, and patient acceptance? If penicillin instead of prayer were being considered, we would not answer the question of use before asking key questions such as these.

Even if it's conceded that prayer and religious behaviors affect health outcomes positively, what then? Should physicians become involved with spirituality? I believe we can decide these questions by means similar to those we have used to approach other sensitive issues in the past. For example, not long ago many physicians believed they should not query patients about their sex lives. Doing so was too personal and disrespectful of privacy. Then the epidemic of sexually transmitted diseases and AIDS arose, and overnight physicians began to see the issue differently. As a result, most physicians have learned to inquire about their patients' sexual behaviors with respect and sensitivity. Inquiries into peoples' spiritual and religious practices can be done with comparable delicacy. Codes of ethics and conduct already exist among hospital chaplains that prohibit evangelization, proselytizing, heavy-handedness, and crass intrusiveness, and similar guidelines can help physicians navigate this territory. Indeed, this is already taking place, as medical students around the country are learning to take spiritual histories from patients in ways that honor privacy and personal choice.[25]

Moreover, consultation is always an option, and physicians can refer patients who voice spiritual concerns to a religious professional. That said, physicians who are not comfortable with spiritual inquiry may sit on the sidelines.

No one expects physicians to be as expert as clergy in these matters, but that does not mean we cannot develop a basic level of expertise. We teach laypersons basic CPR without expecting them to be cardiologists or heart surgeons; just so, physicians can learn the rudiments of spiritual inquiry without becoming as skilled as clergy or hospital chaplains.

This area can also be viewed as a matter of public education. Physicians routinely convey to patients the facts surrounding smoking, the use of seat belts, and protected sex. They can also matter-of-factly deliver information about the latest findings on spirituality and health, and encourage patients to make their own choices in these matters.

Sensitivity and delicacy are eminently achievable if physicians remain patient centered. An internist friend of mine became interested in the prayer-and-healing studies and eventually decided that he had an obligation to pray for his patients. He developed a three-sentence handout that his receptionist gave to each patient as they entered the waiting room. It simply said, "I have reviewed the evidence surrounding prayer and health, and I believe that prayer might be of benefit to you. As your physician, I choose to pray for you. However, if you are uncomfortable with this, sign this sheet below, return it to the receptionist, and I will not add you to my prayer list." Over many years, no one signed the sheet.

Researchers are currently exploring hypotheses from several areas of science that are cordial to the remote effects of prayer and intentionality.[26]

As a theoretical framework gradually emerges, spirituality and the remote effects of healing will begin to seem less foreign, and future physicians may well wonder why we experienced such indigestion over these issues.

The game is early; this field of research hardly existed a few years ago. It took the British Navy around 200 years to require the use of citrus fruit in preventing scurvy aboard its ships, in spite of overwhelming evidence of its effectiveness. The idea that a mere teaspoonful of lime juice a day could prevent such a lethal disease was considered lunacy: theoretical implausibility writ large.

Where spirituality is concerned, let's hope we won't be as obstinate.[27]

A LOOK AHEAD

We cannot know what form our knowledge of consciousness will eventually take, but gone forever, it seems, are the conventional views that identify consciousness solely with the workings of the brain, and which confine the displays of consciousness to the here and now. Why? The reason is, quite simply, that these conventional views cannot account for the Akashic phenomenon, and other ways in which consciousness is manifest in the world. The old models are not adequate for what we see and experience and for what science is demonstrating. That's why the old models may be superseded by the hypothesis described in this book.

Are we adequate to the task of understanding consciousness and its role in the universe? Can we know the mind *with* the mind? William James said:

> I firmly disbelieve, myself, that our human experience is the highest form of experience extant in the universe. I believe rather that we stand in much the same relation to the whole of the universe as our canine and feline pets do to the whole of human life. They inhabit our drawing-rooms and libraries. They take part in scenes of whose significance they have no inkling. They are merely tangent to curves of history the beginnings and ends and forms of which pass wholly beyond their ken. So we are tangents to the wider life of things.[28]

James was a proponent of a "pluralistic universe" so vast, so rich with possibilities, mysteries, and surprises, that he believed we mortals could never fathom it completely.[29] But it's our nature to try.

Acknowledging our limitations is not a concession or admission of defeat. After all, it's the human journey, not the destination, that's most important. As Browning said, "Ah, but a man's reach should exceed his grasp, or what's a heaven for?"[30]

Or—in light of the Akashic and related phenomena—we may opt for an alternative view based on the nonlocal nature of our consciousness, its infinitude in space and time. This is the view of completeness, of "alreadyness" and "nowness," the realization of Tat tvam asi, "Thou art that." This awareness involves the knowledge that the journey has already been run and the destination attained. Our goal is to wake up to this realization. Thus Wittgenstein's magnificent statement: "If we take eternity to mean not infinite temporal duration but timelessness, then eternal life belongs to those who live in the present."[31]

SUMMING UP

Science and the Akashic Experience

This collection of firsthand reports on the Akashic experience covers a great deal of ground. It covers the lived aspect of the experience as well as its practical utility. It reviews current research designed to illuminate the nature and the roots of the experience and provides a review and assessment of what this experience is and how we can think about it. It testifies that the Akashic experience occurs for a wide variety of people and comes in a great variety of forms. Although for purposes of social science documentation the experience of twenty (including the author, twenty-one) individuals is not an adequate basis for evaluation—the sample size is too small—in the context of this widely neglected (and if not neglected then contested) kind of experience, it is significant. It offers evidence that the Akashic experience is not confined to mystics, psychics, shamans, and gurus. It is encountered by people with a wide variety of backgrounds and great diversity of interests.

RECOGNITION OF THE EXPERIENCE

If the Akashic experience is so widespread, why is it not more widely known? The reason is not difficult to find and has been cited in several of the reports already. In the perspective of the modern materialist mentality, the Akashic experience is strange, so it is willingly dismissed, or else relegated to a category people view as esoteric, spiritual, or New Age. Modern people not only dismiss the experience when recounted by others but also repress it if and when it happens to them.

It's noteworthy, at the same time, that when the reality of the Akashic experience is affirmed, many people have an "Aha" experience: *So what I have experienced at one time or another in my life is not just imagination after all.* This, at any rate, is what I have found in the fifteen years since I published the theory of an experientially accessible information and memory field in nature and began speaking about it in lectures and seminars. Apparently, what is needed to lift the Akashic experience to the level of conscious recognition is to produce a credible explanation for it.

For the modern mind a credible explanation is a scientific explanation. Thus, as I wrote in the introduction, I now explore the possibility of connecting the phenomenon of the Akashic experience with the theories and concepts of science, taking account of the latest findings. (Readers interested in delving deeper into the relation of this experience to quantum physics, cosmology, biology, brain and consciousness research, and related disciplines are referred to my previous books.[1])

THE ELEMENTS OF A SCIENTIFIC EXPLANATION

Is there a bona fide scientific explanation for the Akashic experience? Can the lived and empirically tried and tested aspect of this experience be connected in a credible way with what science tells us about the nature of "objective" reality?

I suggest that a scientific explanation is entirely possible. Work at the frontiers of quantum physics, quantum biology, and quantum brain research shows that the brain is physically capable of giving rise to experiences based on information that comes from the external world without having been conveyed by the body's exteroceptive senses. This finding is new and at first sight surprising. Yet there is solid evidence for it. The human brain, with its stupendously complex and coordinated system of neurons, is not merely a classical biochemical system. It is also, and most remarkably, a "macroscopic quantum system"—that is, a system that in some respects acts like systems of microparticles (so-called quanta) even though it's of macroscopic dimension. This finding throws serious doubt on the classical tenet that every extra- or non-sensory experience must be pure fantasy.

The classical tenet can be stated as follows. The brain is a biochemical system that receives and sends information in the form of impulses traveling through the nervous system. In this concept information about the external world is conveyed to consciousness through, and only through, the organs that register external stimuli, namely eye, ear, nose, palate, and skin. Every thought, intuition, image, or experience that is not clearly and evidently conveyed by these sensory receptors must be fantasy based on a recombination of the sense-perceived elements.

The scientifically valid reason why this tenet is no longer convincing is the above-noted finding that the brain is (also) a macroscopic quantum system. The brain carries out functions and processes that were previously thought to be limited to the sub-microscopic world of the quantum. As we shall see below, there are structures in the brain that are of nearly quantum dimension, and these structures receive and send information in the so-called quantum-resonance mode. This is a quasi-instant, multidimensional form of processing and transmitting information that is a basic feature of the life functions of all biological organisms. This mode is clearly recognized in the physical sciences: it is "nonlocality."

Nonlocality contradicts another tenet of the paradigm that still dominates the modern world and has long dominated science: "local realism." Local realism is basic to the modern commonsense concept of the world. It has two main elements: a locality assumption and a reality assumption. The *locality assumption* is that physical effects propagate over space at finite speed and diminish and ultimately disappear with distance. The *reality assumption,* in turn, is that all things in the real world have values and characteristics that are intrinsic to them, rather than being created by their relations or by their observation.

Neither of these assumptions holds in regard to nonlocality. Nonlocality means that physical effects do not propagate over space at finite speeds but spread instantly (or at any rate faster than measurable by existing instruments); and that the characteristics of things, such as the state of particles, is not intrinsic to them but is linked with, and in a sense is created by, the state of other things. And these states may also be determined by our acts of observation.

These are surprising findings, yet they are accepted in contemporary physics. Repeated laboratory experiments show that particles that at any time occupied the same quantum state remain correlated with one another over all finite times and distances. Changes in the state of one of the particles instantly result in changes in the other particle, even when they are no longer connected in any conventional way. Spatial separation, and separation in time, prove to be irrelevant to the correlation of their state: the particles can be anywhere, and can exist, or could have existed, at any time. Space- and time-transcending correlation indicates the physical reality of nonlocality—the kind of "connection at a distance" Einstein dubbed "spooky" (and never became fully reconciled to).

Space- and time-transcending correlation (which Erwin Schrödinger called "entanglement") occurs when the particles—the smallest measurable units of the physical world known as *quanta*—are in coherent states. In their pristine state, prior to any interaction, quanta are indeed in such states. However, when quanta are subjected to interaction (and measurement itself constitutes an interaction—and possibly even observation), they become decoherent, that is, they assume the characteristics of ordinary, "local" rather than "nonlocal," objects. According to classical quantum theory, objects in the everyday world are constantly subjected to interaction and hence they are in a permanently decoherent state. But, as it has turned out, this is not necessarily the case. By the turn of the twenty-first century scientists could correlate the quantum state of entire, seemingly noncommunicating, atoms over many kilometers. And in recent years space- and time-transcending correlations have been discovered in living organisms as well.

It appears that interactions in the body, even those that maintain the body temperature of warm-blooded species, do not destroy the coherence of living organisms. This was a surprise, for previously physicists thought that the Brownian motion (random movement) of particles in the body made them decoherent and thus incapable of "entanglement" over space and time. But recent research (by Kitaev and Pitkanen, among others) shows that the problem of "heat-decoherence" is not insuperable.[2] Networks of quantum particles organized in a specific way (for example, by "weaving" or "braiding" the component particles) appear to be sufficiently robust to maintain quantum coherence even at ordinary temperatures. As Parsons noted, "braiding is robust: just as a passing gust of wind may ruffle your shoelaces but won't untie them, data stored on a quantum braid can survive all kinds of disturbance."[3]

Although the final word has not been spoken, the fundamental divide between the microworld of the quantum and the world of macroscale objects appears to have been breached.

Quantum Correlation and the Organism

In the living organism, quantum effects are not only theoretically possible but also are essential for maintaining the processes of life. The staggering numbers of chemical and physical reactions taking place in the organism are not likely to be coordinated purely by limited and relatively slow biochemical signal transmission. One of the most basic functions of cells—their communication with other cells in the body—has been shown to involve information transmitted through quantum effects: it involves more information, and faster transmission, than any conventional form could account for.

Through quantum effects, cells create a coherent field of information throughout the body. This "biofield" supplements the ordinary flow of information with the multidimensional quasi-instant information needed to ensure the coordinated functioning of the whole organism.

The quantum effects of the biofield are not confined to the physical bounds of the organism: they extend into the environment. Through its biofield, the living organism interacts with all the fields that surround it. Thanks to this interaction the organism is in constant communication with its environment. Because this communication involves quantum effects, the organism is in communication with more than its immediate environment: it is in communication with other organisms whether near or far. In the final count it is connected with the entire sphere of life.

Like sensory information, the information reaching the body through quantum effects originates in the real world and maps events and conditions in that world. Although this kind of spontaneous, extrasensory information is typically denied in the modern world, it was known to traditional peoples—shamans, medicine persons, prophets, and spiritual leaders. Even today, a person of high sensitivity—a "mystic," a "medium," or an unusually intuitive ordinary person—is aware of receiving information of a nonsensory kind. Obtaining this kind of information is not necessarily a delusion, for it is not necessarily *created* by the brain—it might be that it is only *transmitted* by it. This is a fundamental difference. Brain-*created* information could be fantasy, whereas brain-*transmitted* information has its origins in the real world.

How Information Is Conserved in the Universe

We now look at the way information may be conserved in nature. Here we enter terrain that was familiar to traditional wisdom cultures but is new to modern science.

We begin by noting that the information that reaches the mind in an extra- or non-sensory mode does not appear to have conventional limits in space and time. Such information could have come from anywhere, and could have originated at any time in the past. This suggests that the information is not local but universal. It is distributed information in a field that is present throughout nature.

This is a new and perhaps surprising hypothesis, but it's borne out by cutting-edge physics and cosmology. A universal information- and memory-field could exist in nature, associated with the fundamental element of physical reality physicists call the unified field. The unified field, as noted in the introduction, is the originating ground of all the fields, forces, and energies of the universe. It is logical to assume that it not only conserves and transmits energy but also records and conveys *information*. Honoring an ancient insight, this is the aspect or dimension of the unified field that I have called the Akashic field.

The most logical and least speculative hypothesis is that the Akashic field records, conserves, and conveys information in a holographic mode. We know that holograms created with lasers or ordinary light beams are capable of coding, conserving, and conveying a stupendous amount of information in a minimal space—it is said that the entire contents of the Library of Congress in Washington could be coded in a multiplex hologram the size of a cube of sugar. Functioning in a holographic mode, nature's information field could record and convey information on all the things that take place in space and time, from the big bang (or perhaps before) until the big crunch (and perhaps beyond).

We can specify how this information coding would work. We know that every moving object emits quanta of energy, and these quanta carry information on the objects that emitted them. The quanta form coherent waves that propagate in space, and—since space is not an empty domain but a complex field—the waves propagate in the unified field. The expanding wavefronts in the field interact and create specific patterns. These wave interference patterns are similar to the patterns created by interacting beams of light in ordinary holograms—they can be modeled by the same mathematics. This is important, for we know that in holograms the nodes of the interference patterns conserve information on the things and processes that created the interfering light beams, or informed them between their emission and their reception.

In the unified field waves are created by waves of quanta and not waves of light. The former are fully coherent and nonlocal, they "entangle" throughout the field. Thus the resulting interference patterns are quantum—and not ordinary—holograms.

Mathematical physicist Walter Schempp has shown that quantum holograms are coherent and they entangle with one another, just like the vortices that appear in supercooled helium. At extremely low temperatures—below 2.172 on the Kelvin scale—helium becomes a superconducting medium: things move through it without any friction. As reported among others by Russell Donnelly of the University of Oregon, vortices appear in the superconducting medium known as helium-II, and they spread throughout that medium.[4] The vortices entangle throughout helium-II: what happens to one of the vortices has an immediate effect on all the others.

The Akashic field is a field of quantum holograms, a kind of superconducting cosmic medium. There is nothing in that field that could impede the frictionless spread and entanglement of the holograms that arise in it. The quantum holograms created by the waves emitted by objects in space and time entangle throughout the field—that is, throughout space and time. They produce sequences of interference patterns, culminating in the superhologram that is the integration of all other holograms. The superhologram carries information on all the things that exist, and have ever existed: it's the "hologram of the universe."

Receiving Quantum-Hologram Information in the Brain

The Process

There is evidence that the human brain can exchange information with holograms in the environing fields. As Karl Pribram's holonomic brain theory specifies, the brain's receptor and memory functions operate essentially in a holographic mode.[5] Schempp also affirmed that "the conditions which make quantum holography possible are ideally suited to the hypothesis that the brain works . . . by quantum holography."[6]

The process through which the brain can exchange information with holograms in the fields that surround the organism is "phase conjugation," more exactly, "phase-conjugate quantum resonance." This means that the phase of the wavefronts of a hologram in the field is synchronized with the phase of the holographic receptors in the brain, bringing about the resonance that enables the effective transmission of information from the hologram to the brain.

The Structures

The physiological structures that receive and process quantum information in the brain are structures at the supersmall scale. These structures are part of the so-called cytoskeleton. Proteins in the cytoskeleton are organized into a network of microtubules connected to each other structurally by protein links and functionally by gap junctions.[7] The microtubular network is a subcellular network of vastly more elements than the neuronal network. The human brain has approximately 10^{11} neurons, and it has 10^{18} microtubules, so the microtubular network has not ten or a hundred, but 10,000,000 times more elements than the network of neurons. With filaments that are just 5 to 6 nanometers in diameter—the so-called microtrabecular lattice—the microtubular network functions close to the quantum scale.[8]

The cytoskeleton provides structural support and forms a medium of transportation for subcellular materials in the brain and throughout the organism. Previously it was thought to have a purely structural role, but recent evidence indicates that it also conveys signals and processes information.

Physicist Roger Penrose and neurophysiologist Stuart Hameroff suggested that all consciousness emerges from the cytoskeletal structures.[9] Neuroscientist Ede Frecska and anthropologist Luis Eduardo Luna affirmed that the cytoskeleton's ultramicroscopic networks are the most likely structures to perform the computations that transform quantum-level signals into information in the brain. In the emerging consensus, the network of microtubules, with its quantum-scale lattice structure, is the physiological basis of quantum holography in our brain.

The Two Modes of Perception

Given that quantum holography is effectively performed in the brain, there are two kinds of perceptions of the world available to us and not just one. In addition to perceiving the external world through our senses, we can also perceive some of its aspects and elements nonlocally. This mode of perception, called the "direct-intuitive-nonlocal" mode by

Frecska and Luna, is just as real as the "perceptual-cognitive-symbolic" mode of ordinary sensory perception.[10]

Basically the same conclusion was reached by Stanislav Grof. As he wrote in his chapter above, "My observations indicate that we can obtain information about the universe in two radically different ways. Besides the conventional possibility of learning through sensory perception, and analysis and synthesis of the data, we can also find out about various aspects of the world by direct identification with them in an altered state of consciousness." Moreover, as Frecska and Luna asserted, the information accessed in the direct-intuitive-nonlocal mode may be "immense enough to contain holographic information about the whole universe via nonlocal interaction." This meshes with Grof's assertion that "each of us . . . seems to be a microcosm that also has access to the information about the entire macrocosm."

Reading Quantum Holograms in the Akashic Field

If all things create waves in the unified field and interfering waves create quantum holograms in it, in principle our brain can receive information on some aspects or elements of all the things and events in the universe. However, it is clear that we don't receive information on everything in the universe at the same time, nor with the same degree of facility. There must be degrees of access to the information stored in the Akashic field. This limitation is similar to when we access information on the Internet: we don't get everything at once, but only that for which we have the code—for which we have entered a URL. In regard to information in the Akashic field, the cytoskeletal fine structures of our individual brain provide the access code. When these structures are synchronized with a quantum hologram in the field, access is privileged: brain and hologram enter into phase-conjugate quantum resonance.

It is further evident that there are various degrees of access to phase-conjugate information exchange with the Akashic field. The brain is most likely to enter into quantum resonance with the hologram it has itself created: this is the most privileged access. Deriving information from the hologram of one's own brain means accessing the traces it has created in the field. This is the physical basis of long-term memory, including the stupendously detailed and complete memory store that comes to light in the "life review" reported by many people in a near-death experience.

Other people's holograms can also be accessed, especially if they are physically related, as identical twins, or emotionally close, as lovers and friends. When the brain of one person enters into phase-conjugate resonance with the quantum hologram of another's brain, he or she can intuit some elements or aspects of that person's consciousness. (We should note that such intuitions do not necessarily reach conscious awareness: as with all elements of human experience, they can be ignored or repressed.)

Experiments testify that several people can communicate through the Akashic field at the same time. When the brain of one person enters into phase-conjugate resonance with the brain of another, a subtle nonlocal form of communication is created between them. This can embrace an entire group of people, creating what Bache calls the group mind. Then the members of the group may report the kind of experiences Bache has noted above: shared dreams, common hunches, related insights.

Conscious intention can make spontaneous communication between individuals produce specific effects. For example, some natural healers heal from a distance by sending what they call "healing energy" to their patients. In these cases we can assume that the healer's brain was able to enter a relation of quantum resonance with the brain and body of the healee.

The effectiveness of remote healing has been analyzed in statistical studies involving hundreds of cases and has been established beyond reasonable doubt. In turn, the physical basis of the healing process has been tested by measuring the electrical activity of the brain of both healer and healee. As Maria Sagi's report (in part 2) of her remote-healing experiment has shown, the EEG waves appearing in the brain of the healer are reproduced with a delay of a few seconds in the brain of the healee. These patterns occur in the lower regions of the EEG spectrum: in the alpha or even the delta region. Since these regions show activity only in deep sleep or in highly relaxed meditative states, it appears that the human brain enters into phase-conjugate quantum resonance most readily when it is in a deeply altered state.

When two or more persons have a close rapport with each other, their brains resonate more often and more thoroughly. In parapsychology laboratories test subjects are often asked to interact before the experiment to create an empathetic connection with each other. Then one subject is subjected to some form of stimulation—exposed to flashes of light or to weak electric shocks—and the other is not. Yet the pattern of brain activity triggered by the stimulation shows up in the EEG not only of the stimulated subject but also of the other person, even when the latter is beyond the range of sensory contact. Love and profound goodwill increase the frequency and the depth of the entanglement. Jointly entering deep meditation works as well. In experiments conducted by Dr. Montecucco—experiments witnessed by this writer—twelve persons in deep meditation and without sensory contact achieved more than 90 percent synchronization of their EEG waves.

Nonlocal communication has been tested by functional magnetic resonance imaging (fMRI) as well. In a landmark experiment carried out by Jeanne Achterberg and colleagues in Hawaii, eleven healers chose persons with whom they felt an empathic connection.[11] Each recipient was placed in an MRI scanner isolated from sensory contact with the healer. The healers sent energy, prayer, or good intentions—so-called distant intentionality—at intervals that were random and unknown to the recipients. Significant differences were found between the "send" and "no send" (control) periods in the activity of various parts of the recipients' brain—namely in the anterior and middle cingulate areas, precuneus, and the frontal areas. The probability that this correlation between healer input and brain function would have come about purely by chance was calculated as 1 in 10,000.

THE PUZZLING VARIETIES OF THE AKASHIC EXPERIENCE

Phase-conjugate quantum resonance between the human brain and the information stored in the unified field offers a promising foundation for a scientific explanation of the standard varieties of the Akashic experience. Ordinarily we can access quantum holograms in the field because our brain can function in the quantum mode. But what about the Akashic experiences of people whose brain is temporarily nonfunctional? Or cases where the experiencing subject is actually dead? As the near-death, out-of-body, and after-death communication experiences reported in this book testify (see the reports by van Lommel, Brune, and Grof), communication can take place even under such unlikely circumstances.

In near-death experiences (NDEs) the brain is temporarily inoperative; in out-of-body experiences (OBEs) it's in a special state in which it is no longer the physical locus of consciousness; and in after-death communications (ADCs including ITCs, their instrumental transcommunication variety) the brain is permanently out of commission. Yet conscious perceptions, and in some cases also two-way communication, can take place. In NDEs some people have clear and lucid perceptions that can be verified as perceptions of their surroundings; in OBEs they report verifiable perceptions that come from beyond (usually above) the body; and in ADCs defunct individuals seem capable of receiving as well as sending information from beyond the grave.

Before closing I should mention an instrumental transcommunication (ITC) experiment I not only witnessed but participated in myself. I spoke with seemingly disincarnate voices coming through an old-fashioned radio in the complete absence of the usual electromagnetic frequencies and their electronic modulation. I even conversed in my native Hungarian (one of the voices said, "Here we speak all languages"), although the sensitive who conveyed the radio contact with my presumably deceased dialogue partners didn't speak it.

Near-death, out-of-body, and after-death experiences are a puzzling variety of Akashic experience. They are puzzling for modern common sense, and puzzling even for cutting-edge science. They were not puzzling for traditional cultures, where the altered states of consciousness in which they occur were widely known and purposively fostered. But now this puzzling variety of experience is being subjected to observation and testing, and may yield new insights.

In *Quantum Shift in the Global Brain,* I noted that transcommunication occurs through nonlocal quantum resonance between holograms persisting in the Akashic field and the brain of the experimenter. An exchange of information takes place when the experimenter's brain becomes tuned to (that is, enters a relation of phase-conjugate quantum resonance with) the quantum hologram that carries the experiences of the deceased person. The experimenter does not *produce* the voices that come through the radio, the television set, or another electronic instrument (as some hypotheses would have it); he or she merely *transmits* them. A radio tuned to empty regions of the shortwave band, the same as a TV set tuned to empty regions of the broadcast band, is a system in a state of chaos; it produces random static. In this condition it's ultrasensitive, and it's conceivable that it can transform the impulses arriving from the experimenters' brain and nervous system into the audible range of the spectrum of sound waves.[12]

The Akashic field contains not merely a passive record of a person's consciousness, created during that person's lifetime and then persisting unchanged, but also harbors a dynamic bundle of information based on the experiences accumulated in that lifetime. Under suitable conditions this bundle of information—a quantum hologram—is capable of development even in the absence of the living brain and body that created it. I concluded that "given the theoretical tools, the mathematics, and the electronic simulation methods at our disposal it should be possible to discover how complex sets of coherent elements within an information-rich, extremely complex field can function with a form and level of autonomy that permits creating fresh information based on the information already given."[13]

Discovering how this autonomous creativity may be possible is a major challenge for science. Yet it is not unreasonable to expect that it will be cleared up, given the great strides already made in understanding the modalities of information processing in complex systems. Some promising hypotheses have been put forward. Stuart Hameroff, for example, has suggested that "when the metabolism [i.e., the living body] . . . is lost, the quantum information leaks out to the spacetime geometry in the universe at large. Being holographic and entangled, it doesn't dissipate. Hence consciousness (or dream-like subconsciousness) can persist."[14] Such consciousness can indeed persist, given that there is a field in nature that records information on all that happens, and has ever happened, in space and time.

Seeking a credible explanation of the Akashic experience even when it takes place in the absence of a functional brain doesn't call for giving up on science and turning to esoterics and metaphysics. It calls, I believe, for sustained research in the pertinent fields of cutting-edge science, first of all in the physics of quantum holograms and nonlocal coherence as well as the theory of self-regulation and evolution in complex systems.

For mainstream science the Akashic experience is puzzling and vexing; it shouldn't even exist. But the expanding horizons of scientific research are more accomplished and accommodating. Here experimental findings, hypotheses, and theories come to light that offer a solid ground for recognizing the reality of the Akashic experience in all its forms.

The Akashic experience, I conclude, is a veridical extra- or nonsensory experience, a rediscovery and revalidation of the spontaneous insights and intuitions that have accompanied and inspired human culture and consciousness throughout the centuries and millennia of our history.

POSTSCRIPT

A Note on My Akashic Experience

My own Akashic experience has been of a more modest kind than the majority of the experiences reported in this volume, but it has been remarkably constant and enlightening. I began having this experience while I was still a professional concert pianist. It came while I was playing the piano.

Until the fall of 1966, I made my living on the concert stage. Although being a professional musician is a full-time occupation, by about 1959 I developed an intense interest in finding out whatever I could about the fundamental nature of the world, and I devoted all the time I could to pursuing this quest. I had a youthful presumption that with persistence and dedication I could arrive at insights beyond those that had already been attained. I studied all the books I could get my hands on, listened to lectures, and speculated incessantly on the meaning of what I had heard and read.

My mind was particularly free to explore all that emerged in my consciousness while I was playing the piano. As any experienced instrumentalist will tell you, when they have thoroughly absorbed and adequately practiced a piece, performing it doesn't call for conscious effort. Indeed, conscious effort reduces the spontaneity of the performance and renders it dry and mechanical. It's far better to lean back as it were and let the music take wings on its own. Then thoughts, intuitions, and feelings flow freely; one enters what I now recognize as an altered state of consciousness.

Prior to the time of my intensive involvement with the intellectual quest for a deeper knowledge of reality, the altered state I would enter while playing served as a source of inspiration for my musical performance. In this state the hands follow the learned patterns; one is playing the composer's score. But the sense of that pattern, the way to interpret it, comes spontaneously. This feeling is what makes for an inspired interpretation, different from an indifferent one.

Sometimes I had a spontaneous "feel" of the music while playing at home, but more often while performing in public. When all goes well the spirit soars. The public acts as a sounding board, amplifying the feelings that appear in the performer's mind.

When I became deeply involved with the quest to understand the nature of reality, the inflow of musical interpretation was overlaid by an inflow of concepts and ideas. They related to the problems that occupied my mind. Some of the inflowing ideas validated what I had been thinking, and some came with a feeling that I had been mistaken. And some constituted new ideas, fresh insights into the questions with which I was concerned.

Gradually, over a period of years, the "right" ideas cohered into a meaningful concept of the world. This was the concept of an organic and dynamic world, where all things are linked with all other things, and all change and evolve together to higher forms of complexity, harmony, and wholeness.

Evidently, this concept was influenced by my experience of music. There, too, the individual elements lose their separateness and cohere into an integral, harmonious, and meaningful whole. But my world concept was also influenced by reading the works of the philosopher Alfred North Whitehead, especially his major opus, *Process and Reality*. Whitehead's "organic metaphysics" appealed to me greatly; it was meaningful and intuitively correct. This complex metaphysics was not validated for me in its entirety: there were elements of it that didn't mesh with my intuition, such as Whitehead's concept of the Platonic Ideas he called "eternal objects," and some aspects of his concept of the Divine. On the whole, however, the Whiteheadian organic metaphysics was a major factor in my attempt to articulate the concept of a dynamic and integral reality.

My Akashic experiences that began nearly fifty years ago while playing the piano have continued through the years; they continue to this day. Alas, these days I play only occasionally, as my days are filled with other concerns, but there are moments when I enter into the calm reflective state that's conducive to a spontaneous flow of ideas. In the semiconscious white-dream state that precedes waking up fully, a review of the problems that occupied my mind during the day often triggers the intuitive certainty that some of my theories and assumptions are on the right track, while others are not. Fresh ideas appear, suggesting creative solutions to the problems that puzzle me. But the sense that reality is organically whole and dynamically evolving has never varied, nor diminished.

This postscript is not the place to review all the concepts and ideas that have come to me in this intuitive way, but I should mention some that I consider particularly important.

A clear and elementally certain concept has always been the integral unity of the world in all its diversity. The universe is not a layer cake of inherently different kinds of things and processes but a self-evolving system. In the final count there is only one kind of thing that emerges and evolves in space and time, and only one kind of process. Thanks to my friendship with biologist Ludwig von Bertalanffy and physicist Ilya Prigogine and to a study of their theories, I have come to regard the "one kind of thing" as a *natural system*—a complex system that is in constant communication with its environment. The "one kind of process" in the world is that which underlies the evolution—more exactly, the coevolution—of these systems.

A further insight concerned the fundamental elements of physical reality. These, I realized, are energy and information. Matter, on the other hand, is an illusion, created only by our observation of the open systems that coevolve through energy and information.

In the 1970s I learned that in complex systems the evolutionary process is saltatory, punctuated with abrupt shifts—"bifurcations." These are essentially unpredictable but not uncontrollable: the outcome of a bifurcation depends in large measure on the members or components of the bifurcating system. This concept immediately had the flavor of truth. I realized that it's of direct relevance for our time, living as we do at the threshold of a major bifurcation.

The idea that interconnection among open systems in nature is universal and physically real was another basic intuition. The interconnections are intrinsic and mutually constitutive: Whitehead called them "internal relations." They are more fundamental than the causes and effects that relate things and events in the manifest dimensions of space and time. Thanks to this realization, I have never doubted that we humans, remarkably evolved open systems, communicate with each other and with other complex systems in ways that extend beyond the ordinary bounds of space and time. It has not surprised me that such communication has been discovered in discipline after discipline, first of all in quantum physics, cosmology, biology, and consciousness research.

I am, and have been for some time, intuitively certain that everything that has ever happened in the universe remains present in some way; nothing is entirely evanescent. When on a summer evening on the shores of the Mediterranean a friend told me that a scientist I had admired and whose at the time unfinished book I had hoped to read had suddenly passed away, I replied with a certainty that surprised me no less than my friend that his ideas had not passed away: they continued and could be accessed. (This incident, by the way, provided the motivation for me to look seriously into the physics of information conservation in nature.)

Another fundamental intuition that has never ceased to inform my mind is that *psyche* and *physis*, mind and matter, are not separate realities, or even separate elements of the same reality. They are the *same* reality. The difference is in the eye of the beholder: viewed in one way—with one variety of presuppositions—reality appears physical; viewed in another way, it appears mental. In fact, it's both.

I don't wish to leave the impression that all I have been thinking and writing for the past nearly half a century has come to me through such Akashic experiences. Only the basic reality concept has come in this way, but this concept has been decisive: it has functioned as the litmus test for accepting the theories and concepts I encountered. If a concept or theory passed this test, I would take a deeper look at it and attempt to build it into my concept of an integral and dynamic universe. If it didn't, I would not entertain it seriously.

The reality suggested in my Akashic experiences is not an explicitly formulated concept; it's the framework and the foundation for an explicit concept. I have found it a fertile and meaningful ground. I am not alone in being able to build on it; ever more scientists and creative minds are having Akashic experiences. The enterprise of building an explicit and testable concept of the fundamental nature of reality will go on; it's "work in progress." The enterprise is carried forward by people who, like me, are blessed—or cursed—with the consummate passion to find out all they can about life, mind, and universe, and who, like me, have the intuitive certainty that the world of life, mind, and universe is dynamic and whole, and intrinsically knowable.

Notes

Introduction

1. Ervin Laszlo, *Science and the Akashic Field* (Rochester, Vt.: Inner Traditions, 2004, updated second edition 2007).

Chapter 3. Return to Amalfi and the Akashic Home

1. Ervin Laszlo, *Science and the Akashic Field* (Rochester, Vt.: Inner Traditions, 2004, updated second edition 2007).
2. David Loye, *3,000 Years of Love: The Life of Riane Eisler and David Loye* (Carmel, Calif.: Benjamin Franklin Press, 2007).
3. David Loye, *The Sphinx and the Rainbow: Brain, Mind, and Future Vision* (Boston: New Science Library; New York: Bantam New Age, 1984); David Loye, *An Arrow Through Chaos: How We See Into the Future* (Rochester, Vt.: Park Street Press, 2000).
4. Russell Targ and Harold E. Puthoff, *Mind-Reach: Scientists Look at Psychic Ability* (New York: Delacorte Press, 1977).
5. David Loye, *Return to Amalfi* (Carmel, Calif.: Benjamin Franklin Press, 2007).
6. Riane Eisler, *The Chalice and the Blade: Our History, Our Future* (San Francisco: Harper & Row, 1987).
7. Riane Eisler, *Sacred Pleasure: Sex, Myth, and the Politics of the Body* (San Francisco: HarperSanFrancisco, 2007).

Chapter 4. Running with Spotted Fawn in the Akashic Field

1. The author expresses his gratitude to the Saybrook Graduate School's Chair for the Study of Consciousness for its support in the preparation of this essay.

2. Stanley Krippner, "Spotted Fawn's Farewell," in L. Lawson, *Visitations*

from the Afterlife: True Stories of Love and Healing (San Francisco: HarperSanFrancisco, 2000), 57–58.

3. Ervin Laszlo, *Science and the Akashic Field* (Rochester, Vt.: Inner Traditions, 2004).
4. W. G. Roll, "The Psi Field," in W. G. Roll and J. G. Pratt, eds., *Proceedings of the Parapsychological Association, 1957–1964* (Durham, N.C.: Parapsychological Association, 1965), 32–65.
5. D. M. Stokes, "Theoretical Parapsychology," in S. Krippner, ed., *Advances in Parapsychological Research* 4 (Jefferson, N.C.: McFarland, 1987), 77–189.
6. R. Cheney, *Akashic Records: Past Lives and New Directions* (Upland, Calif.: Astara, 1996).

Chapter 6. A Journalist's Encounters with the Akashic Experience

1. Logchempa, *You are the Eyes of the World* (Ithaca, N.Y.: Snow Lion Publications, 2000); Tulku Thondup, *Peaceful Death, Joyful Rebirth* (Boston: Shambhala, 2005); Dzogchen Ponlop, *Mind Beyond Death* (Ithaca, N.Y.: Snow Lion Publications, 2006).
2. Robert Monroe, *Journey Out of the Body* (Garden City, N.J.: Doubleday, 1971); *Far Journey* (New York: Souvenir Press, 1985); *Ultimate Journey* (New York: Broadway Books, 2000); Russell Targ, *Limitless Mind* (Novato, Calif.: New World Library, 2004); Russell Targ, *The End of Suffering* (Charlottesville, Va.: Hampton Roads, 2006).

Chapter 7. The Living Classroom

1. The story of this journey and the corresponding expansion of pedagogical theory and practice is told in my book, *The Living Classroom: Teaching and Collective Consciousness* (Albany: State University of New York Press, 2008).
2. S. Blackman, *Graceful Exits* (Boston: Shambhala Publications, 1997).
3. C. Bache, *Dark Night, Early Dawn* (Albany: State University of New York Press, 2000).
4. R. Sheldrake, *A New Science of Life* (Los Angeles: J. P. Tarcher, 1981); *The Presence of the Past* (New York: Vintage, 1987); *The Rebirth of Nature* (New York: Bantam, 1991).
5. E. Laszlo, *The Interconnected Universe* (Singapore: World Scientific Publishing, 1999); *The Connectivity Hypothesis* (Albany: State University of New York Press, 2003); *Science and the Akashic Field* (Rochester, Vt.: Inner Traditions, 2004).
6. S. Strogatz, *Sync* (New York: Hyperion Books, 2003).

Chapter 10. Visiting the Omniverse Center

1. The following description is an abridged version of a 1994 speech posted online at www.inwardboundvisioning.com/Docs/MONTREALOmniverse Speech.htm.
2. Willis Harman, Oliver Markley, and Russell Rhyne, "The Forecasting of Plausible Alternative Future Histories: Methods, Results, and Educational Policy Implications," *Long-Range Policy Planning in Education* (Paris, Organization for Economic Cooperation and Development, 1973), chapter 3.
3. I later learned that in his 1937 visionary science-fiction classic, *Starmaker,* Olaf Stapledon called this phenomenon "mindedness," through which all individuals of a given species, planet, and so on simultaneously experience themselves as individuals as well as a group consciousness. Stapledon takes the reader on a journey across many different worlds that are comparable in different ways to planet Earth, concluding that mindedness is a likely evolutionary prerequisite for ecological sustainability on a planet whose dominant species is competitive and warlike. Stapledon's *Starmaker* is conveniently available now as a 1968 Dover soft cover edition that also includes his prophetic 1931 novel, *Last and First Men.*
4. This Omniverse Center–based insight led directly to the strongly positive, crisis-transforming scenario contained in the book *Seven Tomorrows: Toward a Voluntary History* by Paul Hawkins, Jay Ogilvy, and Peter Schwartz (New York, Bantam Books, 1982).

Chapter 11. Singing with the Field

1. Raffi Cavoukian and Sharna Olfman, eds., *Child Honouring: How to Turn This World Around* (Westport, Conn.: Homeland Press, 2006).

Chapter 14. Shaping Creative Fields

1. Masahisa Goi (1916–1980), often called "Goi Sensei," was a Japanese philosopher who started an international movement of prayer for world peace. In Japanese, *Sensei* means "teacher." Masahisa Goi was the mentor of the author and became her adoptive father when he designated her as his spiritual successor.
2. For a discussion of the theory of *effect-and-cause*, cf. M. Saionji, *You Are the Universe* (AuthorHouse, 2004); Ervin Laszlo, *You Can Change the World* (SelectBooks, 2003); Masami Saionji, *Vision for the 21st Century* (AuthorHouse, 2005).
3. Regarding useful affirmations, see M. Saionji, *Think Something Wonderful—Exercises in Positive Thinking* (BookSurge, 2005).

Chapter 15. Exploring the Akashic Experience

1. T. Kuhn, *The Structure of Scientific Revolution* (Chicago: University of Chicago, 1970).
2. E. Mitchell, *Psychic Exploration. A Challenge for Science* (New York: G. P. Putnam, 1977).
3. Russell Targ and Harold E. Puthoff, *Mind-Reach: Scientists Look at Psychic Ability* (New York: Delacorte Press, 1977).
4. M. Schlitz and Gruber, "Transcontinental remote viewing," *Journal of Parapsychology* 44 (1980): 305–17; M. Schlitz and Gruber, "Transcontinental remote viewing: A rejudging," *Journal of Parapsychology* 45 (1981): 233–37.
5. M. Schlitz and Haight, "Remote Viewing Revisited: An Intrasubject Replication," *Journal of Parapsychology* 48 (1984): 39–49.
6. M. Schlitz and Honorton, "A ganzfeld ESP study within an artistically gifted population," *Journal of the American Society for Psychical Research* 86 (1992): 83–98.
7. D. Bem and C. Honorton, "Does Psi Exist? Replicable Evidence of an Anomalous Process of Information Transfer," *Psychological Bulletin* 115 (1) (1994): 4–18.
8. W. Braud and M. Schlitz, "Psychokinetic influence on electrodermal activity," *Journal of Parapsychology* 47 (1983): 95–119.
9. M. Schlitz and Braud, "Distant intentionality and healing: Assessing the evidence," *Alternative Therapies* 3 (6) (1997): 62–73.
10. M. Schlitz, D. I. Radin, B. F. Malle, S. Schmidt, J. Utts, and G. L. Yount, "Distant healing intention: Definitions and evolving guidelines for laboratory studies," *Alternative Therapies in Health and Medicine* 9 (3) (2003): A31–A43; M. Schlitz and Durkin, "Compassionate Intention, Prayer, and Distant Healing," *A Self-Paced Learning Program,* DVD (Institute of Noetic Sciences, 2008).
11. M. Schlitz and S. LaBerge, "Covert Observation Increases Skin Conductance in Subjects Unaware of When They Are Being Observed: A Replication," *Journal of Parapsychology* 61 (1997): 185–96; W. Braud and M. Schlitz, "Psychokinetic Influence on Electrodermal Activity," *Journal of Parapsychology* 47 (1983): 95–119.
12. M. Schlitz, S. Schmidt, R. Schneider, J. Utts, and H. Walach, "Distant Intentionality and the Feeling of Being Stared At: Two Meta-analyses," *British Journal of Psychology* 95 (2004): 235–47.
13. R. Wiseman and M. Schlitz, "Experimenter Effects and the Remote Detection of Staring," *Journal of Parapsychology* 61 (1997): 197–207.

14. R. Wiseman and M. Schlitz, "Experimenter Effects and the Remote Detection of Staring: a Replication," *Journal of Parapsychology* 63 (1999): 232–33.
15. M. Schlitz, R. Wiseman, C. Watt, and D. Radin, "Of Two Minds: Skeptic-proponent Collaboration within Parapsychology," *British Journal of Psychology* 97 (3) (2006): 313–22.
16. M. Schlitz, "The Discourse of Controversial Science: The Skeptic-Proponent Debate on Remote Staring," *Journal of Consciousness Studies* 12 (6) (June 2005): 101–5.
17. M. Schlitz, T. Amorok, and C. Vieten, *Living Deeply: Transformational Practices from the World's Wisdom Traditions,* DVD (Oakland, Calif.: New Harbinger, 2008).
18. F. Vaughan, interviewed by Cassandra Vieten and Tina Amorok. Video recording, December 10, 2002, Mill Valley, California.
19. J. Campbell, *The Hero with a Thousand Faces* (Princeton, N. J.: Princeton University Press, 1972).
20. R. N. Remen, interviewed by Marilyn Schlitz. Video recording, May 12, 2003, Mill Valley, California.
21. William James, *The Varieties of Religious Experience* (Cambridge, Mass.: Harvard University Press, 1985).
22. A. H. Maslow, *Religions, Values, and Peak Experiences* (New York: Penguin, 1994).
23. C. G. Jung, *The Structure and Dynamics of the Psyche*, 2d ed. (Princeton, N.J.: Princeton University Press, 1972).
24. M. Schlitz, T. Amorok, and C. Vieten, *Living Deeply: Transformational Practices from the World's Wisdom Traditions,* DVD (Oakland, Calif.: New Harbinger, 2008.
25. W. R. Miller, and J. C'de Baca, *Quantum Change: When Epiphanies and Sudden Insights Transform Ordinary Lives* (New York: The Guilford Press, 2001).
26. R. A. White, "Working Classification of Ehes (Exceptional Human Experiences)," In *Exceptional Human Experience: Background Papers* (Dix Hills, N.Y.: Exceptional Human Experience Network, 1994), 149–50.
27. William James, *The Varieties of Religious Experience* (Cambridge, Mass.: Harvard University Press, 1985), 380.
28. Ibid.
29. M. Schlitz, Amorok, and Micozzi, *Consciousness and Healing: An Integral Approach to Mind Body Medicine* (Amsterdam: Elsevier/Churchill Livingston, 2005).
30. Howard Gardner, *Frames of Mind: The Theory of Multiple Intelligences* (New York: Basic Books, 1983).

31. M. Schlitz, "Worldview Literacy," *Shift Magazine* (Summer 2008).
32. Starhawk, interviewed by Tina Amorok. Video recording, April 25 (Petaluma, Calif.: Institute of Noetic Sciences, 2006).

Chapter 16. Acceding to the Field

1. G. G. Ritchie, *Return from Tomorrow* (Grand Rapids, Mich.: Chosen Books of The Zondervan Corp, 1978).
2. R. A. Moody Jr., *Life after Life* (Covington, Ga.: Mockingbird Books, 1975).
3. D. Kennedy and C. Norman, "What We Don't Know," *Science* 309 (5731) (2005): 75.
4. P. van Lommel, R. van Wees, V. Meyers, and I. Elfferich, "Near-death Experiences in Survivors of Cardiac Arrest: A Prospective Study in the Netherlands," *Lancet* 358 (2001): 2039–45.
5. K. Ring, *Life at Death: A Scientific Investigation of the Near-Death Experience* (New York: Coward, McCann & Geoghegan, 1980).
6. B. Greyson, "Incidence and Correlates of Near-death Experiences in a Cardiac Care Unit," *General Hospital Psychiatry* 25 (2003): 269–76.
7. K. Ring, *Heading Toward Omega: In Search of the Meaning of the Near-Death Experience* (New York: Morrow, 1984).
8. B. Greyson, "Incidence and Correlates of Near-death Experiences in a Cardiac Care Unit," *General Hospital Psychiatry* 25 (2003): 269–76.
9. S. Parnia, D. G. Waller, R. Yeates, and P. Fenwick, "A Qualitative and Quantitative Study of the Incidence, Features and Aetiology of Near Death Experience in Cardiac Arrest Survivors," *Resuscitation* 48 (2001): 149–56.
10. P. Sartori, "The Incidence and Phenomenology of Near-Death Experiences," *Network Review* (Scientific and Medical Network) 90 (2006): 23–25.
11. J. W. De Vries, P. F. A. Bakker, G. H. Visser, J. C. Diephuis, and A. C. Van Huffelen, "Changes in Cerebral Oxygen Uptake and Cerebral Electrical Activity during Defibrillation Threshold Testing," *Anesth Analg* 87 (1998): 16–20; H. Clute and W. J. Levy, "Electroencephalographic Changes during Brief Cardiac Arrest in Humans," *Anesthesiology* 73 (1990): 821–25; T. J. Losasso, D. A. Muzzi, F. B. Meyer, and F. W. Sharbrough, "Electroencephalographic Monitoring of Cerebral Function during Asystole and Successful Cardiopulmonary Resuscitation," *Anesth Analg* 75 (1992): 12–19; S. Parnia and P. Fenwick, "Near-death Experiences in Cardiac Arrest: Visions of a Dying Brain or Visions of a New Science of Consciousness," *Resuscitation* 52 (2002): 5–11.
12. M. Massimini, F. Ferrarelli, R. Huber, S. K. Esser, H. Singh, and G. Tononi,

"Breakdown of Cortical Effective Connectivity during Sleep," *Science* 309 (5744) (2005): 2228–32.

13. P. Van Lommel, "About the Continuity of our Consciousness," *Adv Exp Med Biol* 550 (2004): 115–32; C. Machado and D. A. Shewmon, eds., *Brain Death and Disorders of Consciousness* (New York: Springer, 2004); P. Van Lommel, "Near-Death Experience, Consciousness and the Brain: A New Concept about the Continuity of Our Consciousness Based on Recent Scientific Research on Near-death Experience in Survivors of Cardiac Arrest," *World Futures, The Journal of General Evolution* 62 (2006): 134–51.
14. P. Van Lommel, *Eindeloos Bewustzijn. Een wetenschappelijke visie op de bijna-dood ervaring* (*Endless Consciousness, A Scientific Approach to the Near-Death Experience*). To be published in the English language by Harper Collins in 2010 (Ten Have, Kampen).

Chapter 17. Evidence for the Akashic Field from Modern Consciousness Research

1. J. D. Barrow and F. J. Tipler, *The Anthropic Cosmological Principle* (Oxford: Clarendon Press, 1986); E. Cardeña, S. J. Lynn, and S. Krippner, eds., *Varieties of Anomalous Experience: Examining the Scientific Evidence* (Washington: APA Books, 2000); A. Goswami, *The Self-Aware Universe: How Consciousness Creates the Material World* (Los Angeles, Calif.: J. P. Tarcher, 1995); Bohm, *Wholeness and the Implicate Order* (London: Routledge & Kegan Paul, 1980); K. Pribram, *Languages of the Brain* (Englewood Cliffs, N.J.: Prentice Hall, 1971); L. Prigogine, *From Being to Becoming: Time and Complexity in the Physical Universe* (Gordonsville, Va.: W. H. Freeman & Co., 1980, 1981); R. Sheldrake, *A New Science of Life: The Hypothesis of Formative Causation* (Los Angeles, Calif.: J. P. Tarcher, 1981); I. Stevenson, *Children Who Remember Previous Lives* (Charlottesville, Va.: University of Virginia Press, 1987); I. Stevenson, *Reincarnation and Biology: A Contribution to the Etiology of Birthmarks and Birth Defects* (Westport, Conn.: Praeger, 1997).
2. E. Laszlo, *The Creative Cosmos* (Edinburgh: Floris, 1993); E. Laszlo, *The Connectivity Hypothesis: Foundations of an Integral Science of Quantum, Cosmos, Life, and Consciousness* (Albany: SUNY Press, 2003); E. Laszlo, *Science and the Akashic Field: An Integral Theory of Everything* (Rochester, Vt.: Inner Traditions, 2004).
3. S. Grof, *Beyond the Brain: Birth, Death, and Transcendence in Psychotherapy* (Albany: SUNY Press, 1985); S. Grof, *Psychology of the Future: Lessons from Modern Consciousness Research* (Albany, N.Y.: SUNY Press, 2000).

4. S. Grof and C. Grof, *Spiritual Emergency: When Personal Transformation Becomes a Crisis* (Los Angeles, Calif.: J. P. Tarcher, 1989); C. Grof and S. Grof, *The Stormy Search for the Self: A Guide to Personal Growth Through Transformational Crises* (Los Angeles, Calif.: J. P. Tarcher, 1991).
5. S. Grof, *Beyond the Brain: Birth, Death, and Transcendence in Psychotherapy* (Albany: SUNY Press, 1985); S. Grof, *Psychology of the Future: Lessons from Modern Consciousness Research* (Albany: SUNY Press, 2000).
6. S. Grof, "The Akashic Field and the Dilemmas of Modern Consciousness Research," in Ervin Laszlo, *Science and the Reenchantment of the Cosmos* (Rochester, Vt.: Inner Traditions, 2006).
7. S. Grof, *When the Impossible Happens: Adventures in Non-Ordinary Realities* (Louisville, Colo.: Sounds True, 2006).
8. C. M. Bache, *Lifecycles: Reincarnation and the Web of Life* (New York: Paragon House, 1988).
9. J. G. W. Leibniz, *Monadology* (New York: C. Scribners Sons, 1951).
10. M. Talbot, *The Holographic Universe* (New York: HarperCollins Publishers, 1991).
11. C. G. Jung, *The Archetypes and the Collective Unconscious,* Collected Works, vol. 9, no. 1. Bollingen Series XX (Princeton, N. J.: Princeton University Press, 1959).

Chapter 20. Nonlocal Mind, Healing, and the Akashic Phenomenon

1. L. Dossey, *Recovering the Soul* (New York: Bantam, 1989): 1–20.
2. L. Dossey, "Immortality," *Alternative Therapies in Health and Medicine* 6 (3) (2000): 12–17, 108–15.
3. D. I. Radin, "Unconscious Perception of Future Emotions," *Journal of Consciousness Studies Abstracts* (1996), Tucson II conference, University of Arizona.
4. D. Radin, *Entangled Minds* (New York: Paraview/Simon & Schuster, 2006).
5. S. A. Schwartz, *Opening to the Infinite* (Buda, Tex.: Nemoseen Media, 2007).
6. S. A. Schwartz, *The Secret Vaults of Time* (Charlottesville, Va.: Hampton Roads Publishing Company, 2005).
7. L. Dossey, *The Power of Premonitions.* Forthcoming.
8. D. Lorimer, *Whole in One* (New York: Penguin, 1991); I. Stevenson, *Telepathic Impressions: A Review and Report of Thirty-five New Cases* (Charlottesville: University of Virginia Press, 1970).
9. G. Playfair, *Twin Telepathy* (Walnut Creek, Calif.: Vega, 2003).

10. D. Radin, *Entangled Minds* (New York: Paraview/Simon & Schuster, 2006).

11. L. J. Standish, L. C. Johnson, T. Richards, and L. Kozak, "Evidence of Correlated Functional MRI Signals between Distant Human Brains," *Alternative Therapies in Health and Medicine* 9 (2003): 122–28; L. Standish, L. Kozak, L. C. Johnson, and T. Richards, "Electroencephalographic Evidence of Correlated Event-related Signals between the Brains of Spatially and Sensory Isolated Human Subjects," *Journal of Alternative and Complementary Medicine* 10 (2) (2004): 307–14; J. Wackerman, C. Seiter, H. Keibel, and H. Walach, "Correlations between Brain Electrical Activities of Two Spatially Separated Human Subjects," *Neuroscience Letters* 336 (2003): 60–64; J. Achterberg, K. Cooke, T. Richards, L. Standish, L. Kozak, and J. Lake, "Evidence for Correlations between Distant Intentionality and Brain Function in Recipients: A Functional Magnetic Resonance Imaging Analysis," *Journal of Alternative and Complementary Medicine* 11 (6) (2005): 965–71.

12. D. Radin, "Time-reversed Human Experience: Experimental Evidence and Implications," BoundaryInstitute.org. Accessed July 31, 2000.

13. Ibid.

14. S. A. Schwartz, "Therapeutic Intent and the Art of Observation," *Subtle Energies and Energy Medicine Journal* 1 (1990): ii–viii.

15. R. Otto, *The Idea of the Holy*, trans. J. W. Harvey (New York: Oxford University Press, 1958), 143.

16. L. Dossey, *Healing Words* (San Francisco: HarperSanFrancisco, 1993), 170–72.

17. W. B. Jonas and C. C. Crawford, *Healing, Intention and Energy Medicine* (New York: Churchill Livingstone, 2003), xv–xix.

18. D. Moher, K. F. Schulz, and D. Altman, CONSORT Group, The CONSORT statement: "Revised recommendations for improving the quality of reports of parallel-group randomized trials," *Journal of the American Medical Association* 285 (2001): 1987–91.

19. J. T. Chibnall, J. M. Jeral, and M. A. Cerullo, "Experiments in distant intercessory prayer: God, science, and the lesson of Massah," *Archives of Internal Medicine* 161 (2001): 2529–36; K. S. Thomson, "The Revival of Experiments in Prayer," *American Scientist* 84 (1996): 532–34.

20. L. Dossey and D. B. Hufford, "Are Prayer Experiments Legitimate? Twenty Criticisms," *Explore* 1 (2005): 109–17.

21. J. Fodor, "The Big Idea," *The Times Literary Supplement* (July 3, 1992): 20.

22. J. Searle, front cover quotation, *Journal of Consciousness Studies* 2 (1995).

23. J. Maddox, "The Unexpected Science to Come," *Scientific American* 281 (1999): 62–67. Available at: http://hera.ph1.uni-koeln.de/~heintzma/Weinberg/Maddox.htm. Accessed April 19, 2005.

24. W. Harris, M. Gowda, J. W. Kolb, et al., "A Randomized, Controlled Trial of the Effects of Remote, Intercessory Prayer on Outcomes in Patients Admitted to the Coronary Care Unit," *Archives of Internal Medicine* 159 (1999): 2273–78; W. S. Harris, and W. L. Isley, "Massah and Mechanisms," letter to the editor, *Archives of Internal Medicine* 162 (2002): 1420.
25. V. S. Sierpina, *Taking a spiritual history? Four Models*, available at: http://atc.utmb.edu/altmed/spirit-cases02.htm, accessed March 7, 2005; H. G. Koenig, *Taking a Spiritual History*, available at: http://jama.ama- assn.org/cgi/content/full/291/23/2881, accessed March 7, 2005.
26. D. I. Radin, "Theory," in D. I. Radin, *The Conscious Universe* (San Francisco: HarperSanFrancisco, 1997), 278–87; D. I. Radin, *Entangled Minds* (San Francisco: HarperSanFrancisco, 2006).
27. D. M. Berwick, "Disseminating Innovations in Health Care," *Journal of the American Medical Association* 289 (2003): 1969–75.
28. W. James, *The Correspondence of William James*, vol. 11, John J. McDermott, ed., with Ignas K. Skrupskelis, Elizabeth Berkeley, and Frederick H. Burkhardt, eds. (Charlottesville: University of Virginia Press, 1992–2004): 143–44.
29. W. James, *The Pluralistic Universe* (Lincoln: University of Nebraska Press, 1996).
30. Robert Browning, "Andrea del Sarto," line 98. Wikiquote.com.
31. L. Wittgenstein, Proposition 6.4311, *Tractatus Logico-Philosophicus*, 2nd edition, trans. David Pears and Brian McGuinness (New York: Routledge Classics, 2001), 85.

Summing Up: Science and the Akashic Experience

1. E. Laszlo, *The Creative Cosmos* (Edinburgh: Floris Books, 1993); E. Laszlo, *The Interconnected Universe* (River Edge, N.J.: World Scientific, 1995); E. Laszlo, *The Whispering Pond* (Rockport, Mass.: Element Books, 1996); E. Laszlo, *The Connectivity Hypothesis* (Albany: SUNY Press, 2003); E. Laszlo, *Science and the Akashic Field* (Rochester, Vt.: Inner Traditions, 2004, updated second edition 2007); E. Laszlo, *Science and the Reenchantment of the Cosmos* (Rochester, Vt.: Inner Traditions, 2006); E. Laszlo, *Quantum Shift in the Global Brain* (Rochester, Vt.: Inner Traditions, 2008); E. Laszlo with Jude Currivan, *Cosmos: A Co-Creator's Guide to the Whole-World* (New York: Hay House, 2008).
2. Alexei Kitaev, "Quantum Error Correction with Imperfect Gates," in *Proceedings of the Third International Conference on Quantum Communication and Measurement*, O. Hirota, A. S. Holevo, and C. M. Caves, eds.

(New York: Plenum Press, 1997); Matti Pitkanen, *Topological Geometrodynamics* (Frome, UK: Lunilever Press, 2006).

3. Paul Parsons, "Dancing the Quantum Dream," *New Scientist* 2431 (2004): 31–34.
4. Russell J. Donnelly, "Quantized Vortex Dynamics and Superfluid Turbulence," *Proceedings of a Workshop held at the Isaac Newton Mathematical institute*, Carlo Barenghi, Russell J. Donnelly and W. F. Vinen, eds. (New York: Springer-Verlag, 2001).
5. Karl Pribram, *Brain and Perception: Holonomy and Structure in Figural Processing,* John M Maceachran Memorial Lecture Series (Mahwah, N.J.: Lawrence Erlbaum, 1991).
6. Walter Schempp, "Quantum Holography and Magnetic Resonance Tomography: An Ensemble Quantum Computing Approach," *Informatica* (Slovenia) 21 (3) (1997).
7. Stuart Hameroff, *Ultimate Computing* (Amsterdam: North Holland Publishing, 1987).
8. A nanometer is 10^{-9} meter, that is, one meter divided by 1,000,000,000: one millionth's part of one millimeter.
9. Roger Penrose, *Shadows of the Mind: A Search for the Missing Science of Consciousness* (Oxford: Oxford University Press, 1996); Stuart Hameroff, "Orchestrated Reduction of Quantum Coherence In Brain Microtubules: A Model For Consciousness?," in Stuart Hameroff and Roger Penrose, *Toward a Science of Consciousness, The First Tucson Discussions and Debates,* S. R. Hameroff, A. W. Kaszniak, and A. C. Scott, eds. (Cambridge, Mass.: MIT Press, 1996).
10. Ede Frecska and Luis Eduardo Luna, "Neuro-Ontological Interpretation of Spiritual Experiences," *Neuropsychopharmacologia Hungarica* 8 (3) (2006).
11. J. Achterberg, K. Cooke, T. Richards, L. Standish, L. Kozak, and J. Lake, "Evidence for Correlations between Distant Intentionality and Brain Function in Recipients: A Functional Magnetic Resonance Imaging Analysis," *Journal of Alternative and Complementary Medicine* 11 (6) (2005).
12. In a letter to the author, Dr. Anabela Cardoso, one of the most respected and serious researchers in this field, wrote that one of the discarnate voices she had recorded said, "We interact with the electronic devices through your brain . . . We only need the SW" (a radio tuned to the shortwave band). She concluded, "In my view, your nonlocality hypothesis is one of the most impressive and most credible of the many theories that have so far been proposed in an attempt to explain ITC phenomena." (Letter dated November 17, 2008.)

13. Ervin Laszlo, *Quantum Shift in the Global Brain* (Rochester, Vt.: Inner Traditions, 2008).
14. Stuart Hameroff, *Ultimate Computing* (Amsterdam: North Holland Publishing, 1987).

Index